Department of Trade and Industry

Energy Paper Number 65

Energy Projections for the UK
Energy Use and Energy-Related Emissions of Carbon Dioxide in the UK, 1995 - 2020

March 1995

LONDON: HMSO

© *Crown copyright 1995*
  Applications for reproduction should be made to HMSO's Copyright Unit
  Second impression 1995

ISBN 0 11 515365 9

ENERGY PAPERS

This publication is the 65th in the series of Energy Papers previously published by the Department of Energy. This paper is the seventh in the series to be published by the Department of Trade and Industry.

The series is primarily intended to create a wider public understanding and discussion of energy matters, though some technical papers appear in it from time to time.

The papers do not necessarily represent Government or Departmental policy.

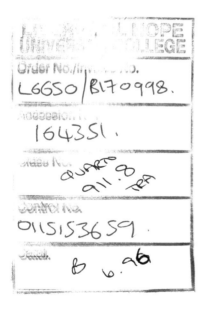

CONTENTS

	Page
Summary	6
1. Introduction	13
2. The Modelling Process	17
The Energy Demand Model	17
Modelling the Electricity Supply Industry	24
3. Assumptions	27
Main Assumptions	28
CO_2 Emissions Assumptions	47
4. Final User Demand Projections	49
Domestic Sector Energy Projections	49
Iron and Steel Sector Energy Projections	54
Other Industry Energy Projections	56
Service Sector Energy Projections	59
Transport Sector Energy Projections	63
Total Final User Energy Projections	65
Total Final User Energy Projections by Fuel	67
Final User Energy Demand Ratios	75
5. The Electricity Supply Industry	82
Background	82
Assumptions	83
Electricity Supply Industry Projections	88
Comparison with EP59	92
6. Primary Energy Demand	93
Background	93
Projections of Primary Energy Demand	93
Primary Energy Ratio	95
7. CO_2 and SO_2 Emissions	97
Background	97
Projected Emissions	97

CONTENTS (continued)

	Page
Annex A Final User Energy Demand by Fuel and Sector	109
Annex B Projections by End Use	122
Annex C ESI Plant Capacity and Fuel Use	129
Annex D Primary Energy Demand	134
Annex E Carbon Dioxide and Sulphur Dioxide Emissions	137
Annex F Energy Demand and Efficiency Trends in the Industrial Sub-Sectors	147
Annex G Very High and Very Low Fuel Prices Energy and CO2 Projections	151
Annex H Final User Energy Demand Price Elasticities	153
Annex I Energy Demand Model Schematics	155
Annex J Glossary	161

TABLES AND CHARTS

		Page
Summary		
	Total Final User Energy Demand	8
	Total Primary Energy Demand	9
	Primary Energy Ratio	10
	UK Carbon Dioxide Emissions 1990 - 2020	10
Chapter 2		
Diagram 2.1	DTI Energy Model Overview	18
Chapter 3		
Table 3.1	Scenario Abbreviations	27
Table 3.2	Long Run GDP Growth Rate Assumptions	28
Table 3.3	Assumed Change in GDP Structure - Central GDP Growth Scenario	29
Table 3.4	Assumed Change in GDP structure - High GDP Growth Scenario	30
Table 3.5	Assumed Change in GDP structure - Low GDP Growth Scenario	30
Table 3.6	Central GDP Growth Scenario - Manufacturing Shares	31
Table 3.7	High GDP Growth Scenario - Manufacturing Shares	32
Table 3.8	Low GDP Growth Scenario - Manufacturing Shares	32
Table 3.9	Crude Oil Price Assumptions - Low and High	34
Table 3.10	Crude Oil Price Assumptions - Very Low/High	34
Table 3.11	Steam Coal Price Assumptions - Low and High	35
Table 3.12	Steam Coal Price Assumptions - Very Low/High	35
Table 3.13	Gas Beach Price Assumptions - Low and High	36
Table 3.14	Gas Beach Price Assumptions - Very Low/High	36
Table 3.15	Crude Oil Prices used by Other Modellers	37
Table 3.16	Steam Coal Prices used by Other Modellers	38
Chart 3.1	Road Transport Fuel Prices	40
Chart 3.2	Heavy Fuel Oil Prices	40
Chart 3.3	Domestic & Service Sectors: Average Price of Petroleum	41
Chart 3.4	Average Domestic Gas Prices	42
Chart 3.5	Average Service Sector Gas Prices	43
Chart 3.6	Average Industrial Sector Gas Prices	44
Chart 3.7	Number of UK Households	45
Table 3.17	Sterling - Dollar Exchange Rate Assumptions	46

TABLES AND CHARTS (continued)

		Page
Table 3.18	Carbon Savings Originally Assumed in the Climate Change Plan	48

Chapter 4

Table 4.1	Domestic Sector Energy Demand Projections	49
Chart 4.1	EP65 Domestic Sector Energy Demand Projections	53
Table 4.2	Iron and Steel Sector Energy Demand Projections	54
Chart 4.2	EP65 Iron & Steel Sector Energy Demand Projections	55
Table 4.3	Other Industry Sector Energy Demand Projections	56
Table 4.4	Other Industry Projections Compared to EP59	58
Table 4.5	Service Sector Energy Demand Projections	59
Chart 4.3	EP65 Service Sector Energy Demand Projections	61
Table 4.6	Transport Sector Energy Demand Projections	63
Table 4.7	Transport Sector Projections Compared to EP59	63
Chart 4.4	EP65 Transport Sector Energy Demand Projections	64
Table 4.8	Total Final User Energy Demand Projections	65
Table 4.9	Final User Sector Growth Rates Per Annum	65
Chart 4.5	EP65 Total Final User Energy Projections	66
Table 4.10	Electricity Demand Projections	67
Table 4.11	Per Annum Electricity Growth Rates EP65 v EP59	68
Chart 4.6	EP65 Electricity Demand Projections	68
Table 4.12	Gas Demand Projections	69
Table 4.13	Per Annum Gas Growth Rates EP65 v EP59	69
Chart 4.7	EP65 Gas Demand Projections	70
Table 4.14	Oil Product Demand Projections	71
Table 4.15	Per Annum Oil Product Growth Rates EP65 v EP59	71
Chart 4.8	EP65 Petroleum Product Demand Projections	72
Table 4.16	Solid Fuel Demand Projections	73
Table 4.17	Per Annum Solid Fuel Growth Rates EP65 v EP59	74
Chart 4.9	EP65 Solid Fuel Demand Projections	74
Chart 4.10	Domestic Sector Energy Ratio	76
Chart 4.11	Iron & Steel Sector Energy Ratio	77
Chart 4.12	Other Industry Sector Energy Ratio	78
Chart 4.13	Service Sector Energy Ratio	79
Chart 4.14	Transport Sector Energy Ratio	80
Chart 4.15	Total Final User Energy Demand/GDP Ratio	80

TABLES AND CHARTS (continued)

		Page
Chapter 5		
Chart 5.1	ESI Capacity, GW, Difference from EP59 CL Scenario	**84**
Table 5.1	Load Factor Restrictions on ESI Plants	**88**
Chapter 6		
Chart 6.1	UK Primary Energy Demand, 1990 - 2020	**94**
Chart 6.2	Primary Energy Ratio 1970 - 2020	**95**
Chapter 7		
Chart 7.1	UK Carbon Dioxide Emissions	**98**
Table 7.1	UK CO2 Emissions in 2000	**99**
Chart 7.2	Difference between CH and CL Emissions	**104**
Chart 7.3	% Reduction in Sulphur Dioxide Emissions from 1980 Base	**107**

SUMMARY

1. This report presents the results of work by the Department of Trade and Industry to update the Government's projections of future energy demand and energy-related carbon dioxide (CO_2) emissions, last published in Energy Paper 59 (EP59) in 1992. The three main uses of the projections are:

- to monitor whether the UK is on course to meet its existing international commitments to limit CO_2 emissions;

- to inform debate on possible future commitments, particularly in the context of the first Conference of Parties to the Framework Convention on Climate Change (FCCC), which will take place in Berlin in April 1995;

- to monitor the general development and direction of energy markets, for example, as a background against which to view the potential future role of nuclear power in the UK.

2. It is important not to underestimate the uncertainties inherent in projecting the future path of energy demand. As with previous projections, rather than rely on a single forecast the projections instead provide a view of the possible future levels and composition of energy demand in the UK in the period to 2020 based on a set of six different scenarios of growth in the economy and of world fossil fuel prices. This provides a range of energy and CO_2 projections.

DEVELOPMENTS SINCE ENERGY PAPER 59

3. Although the projections are primarily based on an analysis of historical trends in energy use and its relationship to factors such as economic growth and fuel prices, they also reflect the effects of existing Government policies on energy and the environment. For example:

- the existing level of effort by the Energy Efficiency Office (EEO) to encourage energy efficiency is assumed to continue;

- the measures in the 1994 Climate Change Programme (CCP) (subject to certain adjustments to take account of subsequent developments, as explained elsewhere in this paper), which are assumed to meet their energy saving targets. The projections estimate that this would save about 8 - 9 MtC (million tonnes of carbon) in 2000;

- the UK's international commitments to reduce sulphur dioxide (SO_2) emissions are assumed to be achieved, in some scenarios through further abatement measures.

The projections do not attempt to incorporate policies or commitments which Governments might adopt in the future.

4. Other developments in energy markets, some of which were only incompletely anticipated in EP59, are also reflected in the projections. Three stand out:

- the rate at which the privatised electricity supply industry (ESI) has built new Combined Cycle Gas Turbine (CCGT) stations has been higher than assumed in EP59; nuclear output has also been greater than expected. This has had the effect of shifting the fuel mix away from coal and is a major reason for the downward revision in future CO_2 emissions;

- the length and depth of the economic recession was not fully anticipated, although growth in the last year or so has gone some way to make up the short-fall compared to EP59;

- there has been a general lowering of expectations for future fossil fuel prices.

5. In the long term, the growth in the demand for energy in some end uses is likely slow down as markets approach saturation (for example, domestic central heating growth is likely to slacken as the majority of dwellings now already have central heating). By adopting a more disaggregated approach in the underlying analysis, these projections are better able to take such effects into account than were their predecessors in EP59. One result of this disaggregation is to allow the projections more accurately to reflect Government measures targeted at particular sectors or

end uses. Another is significantly to reduce the rate of growth of energy demand associated with high rates of economic growth in the long-term.

FINAL ENERGY DEMAND

6. The six projections for total final energy demand combine three different scenarios for economic growth (low, central, high) with two scenarios for fuel prices (low, high). Average growth in demand is expected to be between 0.7% and 1.4% per annum in the period 1990 - 2000. After making allowance for differences in the definition of final demand between EP59 and EP65, this leaves energy demand in 2000 within, but towards the top half, of the range anticipated in EP59. The range of long-run growth rates is expected to be broadly similar to the range over the period 1990 - 2000. At the top end of the range this implies a significantly lower level of energy demand by 2020 than was thought likely in EP59.

Total Final User Energy Demand				Mtoe
Scenario[1]	1990[2]	2000	2010	2020
LL	147	162	176	195
LH	147	158	168	187
CL	147	166	185	210
CH	147	161	178	203
HL	147	169	192	222
HH	147	164	185	215

1: Table 3.1 contains a full description of the six scenarios.
2: 1990 data not temperature adjusted.

PRIMARY ENERGY DEMAND

7. A dominant feature of the projection of the ESI is a shift towards more efficient and cleaner modes of electricity generation (primarily CCGTs), resulting in a lower level of primary electricity demand for any given level of final demand. Largely as a result of this, primary demand is expected to grow more slowly than final demand. However the rate at which gas' share will continue to expand at the expense of other fuels, both in electricity generation and in other markets, will depend on it maintaining

its position as a competitive and secure source of supply. In the long-term, the projections suggest that, given favourable assumptions about future fuel price relativities and about generators' concerns to maintain a diverse portfolio of plant, there could be scope for new generating technologies burning coal to establish themselves.

Total Primary Energy Demand[1]				Mtoe
Scenario	1990	2000	2010	2020
LL	221	231	245	262
LH	221	226	237	253
CL	221	237	257	283
CH	221	232	249	273
HL	221	240	266	298
HH	221	235	258	289
EP59 Range	221	216 - 245	237 - 311	256 - 381

1: temperature adjusted

8. The energy ratio measures the ratio between energy use and output or income. In its most aggregated form, it is based on total primary energy demand and gross domestic product. There has been a long-term trend decline in the energy ratio, reflecting not only improvements in energy efficiency, but also the declining importance of energy intensive industries. This downward trend is expected to continue throughout the period to 2020.

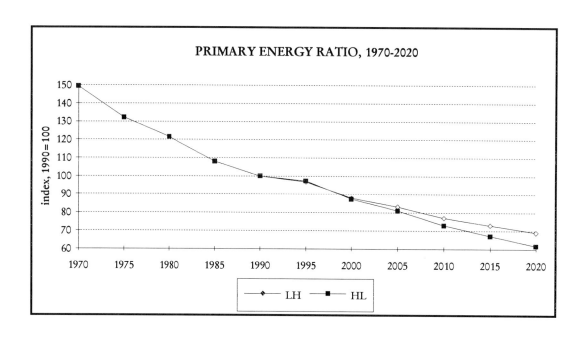

CARBON DIOXIDE EMISSIONS

9.   A view of future energy-related CO_2 emissions can be obtained by applying estimates of the carbon content of individual fuels to the projections of energy demand. The table below shows the CO_2 projections associated with the six energy demand scenarios and compares them to the range in EP59.

UK Carbon Dioxide Emissions, 1990-2020					MtC
Scenario	1990[1]	2000	2005	2010	2020
LL	158	147	157	155	171
LH	158	144	154	154	173
CL	158	150	162	162	184
CH	158	148	159	161	188
HL	158	152	165	167	193
HH	158	151	163	165	197
EP59 Range	158	156 - 178	165 - 200	172 - 227	186 - 283

1: Not temperature corrected

The main features of the CO_2 projection are that:

- on the assumption that the non-fiscal CCP measures (including the Energy Savings Trust (EST)) achieve the full energy savings originally anticipated for them, the CCP might contribute savings of 8 - 9 MtC by 2000. This includes projected savings of 1.6MtC from the EST. However the EST is currently redrawing its plans. It estimates that the programme of activities, which are already established and financed, will save about 0.3 MtC;

- the UK would be likely to meet its CO_2 target of returning emissions in the year 2000 to their 1990 level with a margin of 6 - 13 MtC to spare, even if the EST were to achieve no more than the level of savings from activities already established (it may in practice do better than this);

- during the decade following 2000, CO_2 emissions are expected to grow relatively slowly. By 2010 the projected range is from nearly 5 MtC below to 8 MtC above the 1990 level;

- beyond this, faster growth of CO_2 emissions could reassert itself, especially in the scenarios which show a some recovery in coal demand;

- mainly because the top of the range of likely energy demands has been reduced, the range of CO_2 emissions projected for 2020 is significantly lower than in EP59.

10. Substantial uncertainty, increasing the further into the future one looks, is attached to these projections; this is reflected in the way the range gradually widens over time. However the projections cannot, and do not attempt to, allow for all possible future outcomes. Energy technologies might evolve in new directions, geopolitical developments might substantially alter the prices and security of fuel supplies, social and economic changes might accelerate or reverse the historical trends upon which the projections are based. But, if they are to be of any use for the development of policy on energy and the environment, a balance has to be struck between, on the one hand, concentrating on too narrow a view of the future and, on the other, producing such a wide range of projections that

they convey little useful information. The results reported here should be regarded as a general framework for the discussion and development of policy, but not as an attempt to produce precise forecasts of energy demand and CO_2 emissions.

CHAPTER 1   INTRODUCTION

1.1   Ever since the 1950s official projections of future energy demand have been produced at periodic intervals. While most of the energy industries were in public control, these fulfilled a planning role, but one that has become increasingly inappropriate in recent years as privatisation has shifted the decision making away from state ownership to market forces. The recent sale of British Coal marks another step down this path; only nuclear power now remains in public ownership. However, as the need for the Government to take responsibility for a detailed long term view of energy demand and supply has diminished, concern about the environmental impacts of energy production and consumption has grown. These effects range from the local to the global; and from the transitory to the irreversible. But a common feature is that they are, for the most part, not mediated through the market place, except where Government has intervened through regulation, taxes or other policies. It is mainly in order to develop and evaluate such policies that a view of possible long-term developments in energy supply and demand and of associated pollutants is needed.

1.2   This paper presents the results of work by the Department of Trade and Industry (DTI) to update and revise the projections of future energy demand and carbon dioxide emissions which were published in Energy Paper 59 (EP59) in 1992.[1] Those projections provided a backdrop to the UK's participation in the Earth Summit in Rio in June 1992 and later underpinned the Government's decision to aim to return CO_2 emissions to 1990 levels by the year 2000. In January 1994, the Government published details of its strategy for meeting its commitment, through a range of measures aimed at encouraging energy efficiency across all sectors of the economy.[2]

1.3   Since the projections contained in EP59 were prepared there have been changes in the energy sector and in the wider economy - some of which

---

[1] Energy Related Carbon Emissions in Possible Future Scenarios for the United Kingdom. Energy Paper Number 59 , HMSO, 1992.

[2] Climate Change: the UK Programme. CM2427, HMSO, 1994

were anticipated but some of which were not. With the first Conference of Parties to the Framework Convention on Climate Change Convention due to take place in Berlin in March 1995, the time is now right to review and reassess whether the UK is on track to meet its CO_2 commitment for the year 2000 and to consider whether this commitment should be extended or strengthened beyond then. The new energy and CO_2 projections contained in this document represent an essential building block in this process.

1.4   The analytical models which were use to produce the projections in EP59 have been developed substantially since 1992. Most of the changes have focused on providing a less aggregated view of future energy demands, in order to increase the projections' usefulness as an input to the development of policy on energy and the environment. To this end, both the methods employed and the results are described here in considerably greater detail than in EP59. However, a word of caution is in order: the projections attempt to provide a set of quantified, detailed, views of the future, consistent with specified assumptions, but the further into the future one looks, the greater the uncertainty and the less likely it is that any single forecast of energy demand will turn out to be correct. The approach adopted here, as in EP59, has been to consider a range of possible scenarios for the main determinants of energy demand in the hope that by doing so the demand projections will both encompass the likely range of possible outturns and, as importantly, indicate where the major uncertainties could arise.

1.5   In order to keep the analysis tractable, two main factors - world fossil fuel prices and the UK's economic growth rate - have been chosen for particular attention. Analysis of historic trends in energy demand shows that these two factors play a major determining role, but both are difficult to forecast. Accordingly the projections reported here make use of three different economic growth scenarios (high, central and low) and two fossil fuel price scenarios (high and low), giving a set of six scenarios in all. No single scenario can be viewed as the central one. Nor have subjective probabilities been assigned to the different scenarios, because past experience suggests that such probabilities are generally unreliable and tend to focus attention on one or two scenarios at the expense of the others. Although for some purposes it might be pragmatic to concentrate only on the centre of the range, in most circumstances policy formulation needs to take account of the whole range of projections.

1.6 This still leaves some other areas of uncertainty largely uninvestigated - particularly the role that new technological developments, in either energy supply or demand, could play. The methodology used here implicitly (and in some cases explicitly) allows for continuing technological improvements, but, outside the electricity generating sector, it does not offer a detailed appraisal of the potential contributions of the full range of energy producing and using technologies currently available or under development. Other recent studies have looked more fully at this issue[3].

CURRENT POLICIES ON ENERGY AND THE ENVIRONMENT

1.7 Current Government policies on energy and the environment have been incorporated in the projections wherever possible. These include

- growing liberalisation and competition in energy markets, both within the UK and more generally in Europe and beyond.

- the 1994 CCP measures, as amended by the changes announced in November and December 1994. With the exception of these modifications it has been assumed that the measures save the amounts of energy ascribed to them in the CCP. DoE are currently reviewing progress in achieving these savings;

- emissions standards on new generating plant;

- limits on the output from existing fossil fuel fired stations, in line with Her Majesty's Inspectorate of Pollution (HMIP) authorisations;

- reductions in sulphur dioxide emissions in line with the UK's targets for 2005 and 2010.

---

[3] Energy Technologies for the UK. Energy Paper 61, HMSO, 1994.

THE NUCLEAR REVIEW

1.8   For the purpose of these projections nuclear power is treated in the same way as any other electricity generating technology: decisions to build and operate nuclear stations are based on commercial criteria of cost and performance (subject to the environmental constraints listed above). This gives no special value to nuclear's potential contributions to CO_2 abatement and to fuel diversity. Both issues are being dealt with fully in the context of the Nuclear Review, but two points are worth making here:

- The projections are a useful baseline from which to measure the potential cost-effectiveness of nuclear power as a means of meeting future CO_2 targets.

- Within the framework of competitive markets, both energy consumers and the major energy producers (particularly the generators) will have to weigh up how far they wish to become reliant on any one fuel or source of supply. For the purpose of the projections, it has been sufficient to make a broad judgement about how, in its choice of plant, the electricity supply industry will balance short-term cost minimisation against diversity of supply.

CHAPTER 2. THE MODELLING PROCESS

2.1 The energy scenarios described in this paper are based on a set of interlocking economic models of final user energy sectors and the electricity supply industry. On the basis of assumptions about fossil fuel prices, economic growth and other relevant factors the models can be used to investigate possible scenarios for UK energy prices, demand and supply. The scenarios presented in this paper are based on assumptions which cover a wide range of possible outcomes. They are not forecasts in the sense that Her Majesty's Treasury (HMT) publishes forecasts of GDP and other macro-economic variables. Certainly, no attempt has been made to produce a single most likely outcome. With these qualifications in mind the general structure of the modelling system is discussed below.

THE ENERGY DEMAND MODEL

2.2 The final user energy demand model used for EP59 has been updated for Energy Paper 65 (EP65). The EP65 model that has emerged from this update consists of approximately 130 econometric equations. Of these 130 equations around 60 are fossil fuel share equations, 20 are stock equations and the remaining 50 are energy demand equations. The general structure of the EP65 model is shown in diagram 2.1. Annex I contains model diagrams for the domestic, service, other industry and road transport sectors. In addition a process flow diagram is shown for the iron and steel sector. This process flow diagram provides a good guide to how crude steel output and energy demands are associated in the iron and steel sector model.

2.3 Annex H provides some estimates of the energy model's price elasticities of demand. These estimates provide the reader with a feel for how the energy model responds to higher prices.

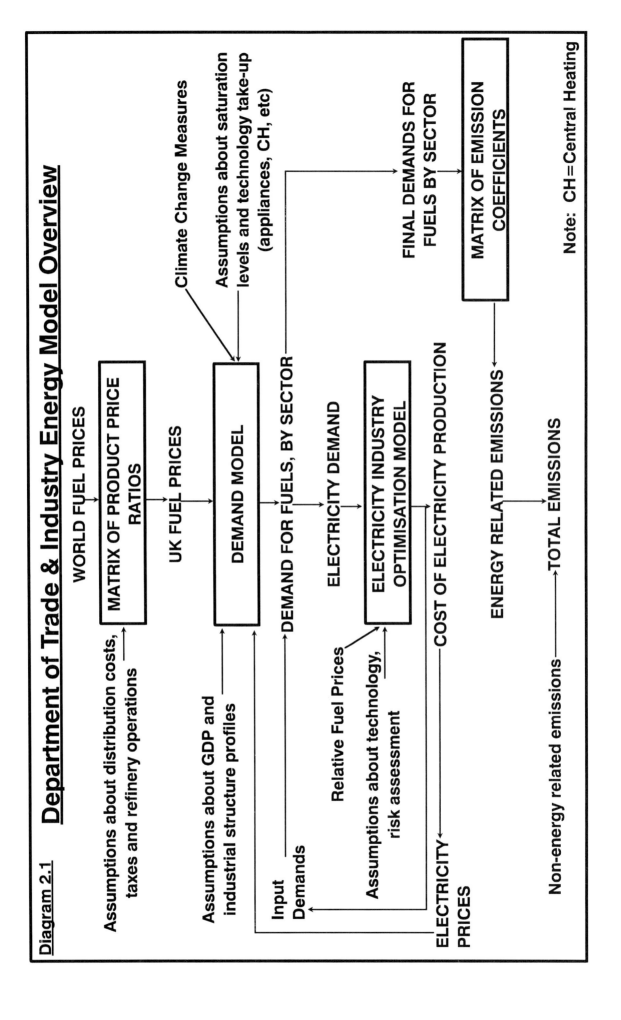

2.4 Energy demand was disaggregated into six categories of final user in EP59. In EP65 this disaggregation has been extended to 13 sectors. This is shown below:

EP65 Energy Demand Sectors

1. Services
2. Domestic
3. Iron & Steel
4. Agriculture
5. Transport
6. Non-ferrous Metals
7. Engineering
8. Mineral Products
9. Chemicals
10. Food, Drink & Tobacco
11. Textiles, Leather & Clothing
12. Paper, Printing and Publishing
13. Construction & Other Industries

EP59 Energy Demand Sectors

1. Services
2. Domestic
3. Iron & Steel
4. Agriculture
5. Transport
6. Other Industry

2.5 The EP59 model aggregated all of the eight manufacturing categories (i.e. sectors 6 to 13 sectors inclusive in EP65) into one single sector. In addition each sector's energy demand has been disaggregated to a much greater degree than was the case for EP59. This greater disaggregation facilitates the comparison between the EP65 results and the results of technology based optimising models (often described collectively as "bottom up" models).[1] The specification and estimation of the econometric equations in EP65 take more account than did EP59 of bottom up data supplied from the Building Research Establishment (BRE) and Energy Technology Support Unit (ETSU).

2.6 The disaggregation used in the model permits sectoral trends to be identified and incorporated into the scenarios. A more aggregated approach would be more likely to miss these subtle changes in the structure of demand over time. Where possible, each of the final user sectors has been further

---

[1] For example, the MARKAL model, used in Energy Paper 61, Energy Technologies for the UK. HMSO, 1994.

disaggregated by end use. Energy efficiency is obviously an important element in determining the demand for energy and is discussed below.

2.7   Each of the final user sectors consists of a suite of econometric equations that attempt to explain past energy demand as a function of other variables such as prices and income or output levels. These equations therefore implicitly incorporate historic trends in efficiency. Where reliable information exists, stock data has been incorporated into the equations (e.g. the stock of cars). To the extent that Government policies, particularly in the area of energy efficiency, have affected these trends the models incorporate these effects and, by implication, the equations assume a continuation of the level of Government support for energy efficiency achieved in recent years.

2.8   The latest version of the Science Policy Research Unit's (SPRU) boiler and combined heat and power (CHP) model is used in the services, other industry and iron and steel sectors as a shares model to divide total boiler and CHP fossil fuel demands into the three different fossil fuels, namely coal, gas and oil. These shares are then applied to projections of total fossil fuel demand in each sector, obtained from econometric equations linking output and fuel prices to demand.

2.9   The SPRU model has also been used to incorporate the Government's target of 5GW of CHP capacity by 2000[2]. Beyond 2000, the SPRU model has also been used to model future CHP capacity.

2.10   The energy demand models used for both EP65 and EP59 are estimated in useful therms rather than delivered therms. Delivered therms are what consumers pay for, e.g. metered consumption of gas. However, because some heat is normally lost at the point of final consumption (e.g. much of the heat from an open domestic coal fire escapes up the chimney) the amount of useful energy obtained from a fuel is less than the delivered amount. The ratio of useful to delivered energy therefore depends in part on how efficiently consumers make use of energy. While it is difficult to be precise about any individual fuel's useful therm conversion factor (useful

---

[2] SPRU developed the boiler and CHP model. Responsibility for the particular results reported here remains with the DTI.

therms/delivered therms) at any point in time, relative useful therm conversion factors between different fuel types are easier to estimate. The relationship between useful therm conversion factors in the DTI model is:

$$\text{Electricity} > \text{Gas} > \text{Oil} > \text{Solid Fuels}$$

Energy Efficiency and Energy Intensity

2.11 The rate of energy efficiency change within each end-use sector is an important component of the change in delivered energy and therefore in emissions.

2.12 Energy efficiency and energy intensity are different concepts. Energy efficiency is the technical efficiency with which a demand for useful energy can be met. If less delivered energy is used to meet a given level of useful energy demand over time, then energy efficiency is said to have increased. Energy intensity relates to the amount of energy used per unit of output produced in a sector or the economy as a whole. In many situations energy efficiency and energy intensity will move together, but this is not always the case. For example with static GDP a shift of output from low energy intensity sectors into higher energy intensity sectors could cause average energy intensity to rise, even though energy efficiency in each sector is unchanged.

Energy Efficiency in the Demand Equations

2.13 The energy demand equations used in EP59 did not easily allow energy efficiency changes to be distinguished from intensity changes. The more detailed disaggregation in EP65 makes it easier to make this distinction, but does not solve the problem completely because, even with the greater degree of disaggregation, the models do not generally identify energy demand at a technology or process specific level. Because of this the rest of this paper uses the term energy efficiency to cover both concepts, except where this usage would be misleading.

2.14 The econometric equations used for calculating energy demand allow directly for changes in energy efficiency in two ways. First, the demand for energy tends, ceteris paribus, to rise less rapidly than economic activity. This implies increases in energy efficiency as GDP grows. This partly reflects the application of new energy efficient technologies, partly the tendency for growth to be concentrated in industries with low energy intensities, e.g. the service sector, and partly the tendency of domestic consumers to spend a declining share of their income on fuel as they become richer.

2.15 Secondly, the demand for energy rises less rapidly for any given level of activity if energy prices rise faster than general inflation. This is due to capital or labour being substituted for energy (e.g. through more insulation) or to research and development expenditures being diverted to energy saving technologies and products. It also arises for two other reasons (not usually considered as energy efficiency per se): with higher energy prices, consumers may desire lower heating and lighting standards and consumption patterns will tend to switch to products that require less energy to manufacture and/or use.

2.16 The choice of whether or not to invest in energy saving technology is affected by the discount rate used for the appraisal. The models used here have been estimated using historical data and therefore implicitly reflect the discount rates or appraisal methods which individuals or companies actually use.

2.17 The econometric estimation of these energy efficiency changes is based on past trends and relationships. Some of the main areas where future trends seem likely to diverge from past trends lie within the domestic and service sectors. One important development in this area is the way in which saturation levels have been built into the model. The discussion below describes how these have been incorporated into the energy model.

2.18 Improvements to energy efficiency in space heating have historically been partly outweighed by the effect of consumers choosing higher comfort levels. As central heating starts to approach saturation levels it is reasonable to assume that the increase in the associated energy consumption will become less rapid. This trend can be seen in the EP65 domestic sector projections, as beyond the year 2000 domestic sector energy demand can be seen to saturate. Stock variables, such as the number of homes with central heating, are used in the EP65 space heating equations to embody these saturation effects. Implicit allowance is also made for the greater uptake of energy efficiency measures, such as double glazing and loft insulation, that will stem from higher real incomes.

2.19   Domestic sector use of electricity has risen rapidly as the ownership of electricity-intensive appliances, such as freezers and tumble dryers, has spread. As a significant number of households now own these electricity-intensive appliances the market for these products is approaching saturation. The future growth in the demand for energy associated with these products is likely to be less than in the past. The model estimates electricity consumption in these domestic appliances by using saturation curves (supplied by the BRE) and average appliance consumption assumptions (supplied by the EEO).

2.20   Apart from the major electrical appliances there is a large number of minor electrical appliances. The demand associated with these appliances is captured via a separate demand equation. A feature of this equation is that, as income levels rise, the incremental effect of higher income on demand diminishes.

2.21   Service sector energy demand would appear, on the basis of growth in the last 20 years, to have output elasticities well above unity (i.e. energy demand growth in this sector has been more rapid than growth in the size of the sector itself). The air conditioning and 'other electricity' output elasticities are particularly large. It seems unlikely that such high elasticities can be sustained in the long-run. For projection purposes air conditioning's electricity demand is assumed eventually to saturate at around two and a half times its current level and 'other electricity' demand is projected in such a way that the eventual long-run output elasticity is +0.65.

2.22   Although one can debate the precise assumptions made about the future growth of the service sector's 'other electricity' demand it seems likely that this growth will moderate in future. Certainly in the US new regulations (Energy Star Programme) require that all personal computers and printers purchased by the US government conform to a specified energy efficiency standard. The widespread availability of such appliances in the US has resulted in similar appliances (at least partially) penetrating the UK market. In addition saturation effects alone will ensure that growth moderates.

MODELLING THE ELECTRICITY SUPPLY INDUSTRY

Background

2.23 In EP59, the electricity supply industry modelling procedure was described briefly in Chapter 2. This did not describe in any detail the nature of the plant or fuel data inputs and it is in these areas where most development work has taken place since EP59 was published. For the EP59 modelling exercise, all coal plants were put into one of four 4 categories - medium coal, intermediate coal, large coal and coal with flue gas desulphurisation (FGD) retrofitted.

2.24 The number of (coal) fuels in EP59 was also limited so that effectively a large number of coal plants were treated as though they each had the same plant characteristics and fuel input prices and thus ran at the same load factor, etc. Whilst the modelling results could not be considered to be dramatically awry, it was felt that the response could be rather too discrete at times - for example, small changes in the relative price of coal and oil could lead to very large changes in the amount of each fuel consumed.

2.25 The level of disaggregation in the model used for EP59 was partly dictated by limitations on computing capacity - some modest increase in the number of plant types could have been carried out, but significant expansion would have increased model run times disproportionately. With the advent of computers with much faster processing times, the modelling capability within the ESI sector has been greatly expanded. These changes interact to a considerable degree and so are described jointly below.

Plant and Fuel Modelling Changes

2.26 All of the UK's major coal power stations are now modelled individually. Each coal station has its own plant characteristics, such as age thermal efficiency, emissions abatement, etc. In addition, each plant has its own supply curve for coal inputs for each year considered. Each station can access deep-mined coal, opencast coal or imported coal. The supply curve is estimated from the assumed pithead or port landing cost of each source to which is added the transport cost to each power station. Each power station will therefore have many coal supply choices from which to choose. The model then allocates fuels to each power station in such a way as to minimise the overall system cost, subject to meeting any other constraints placed upon it (e.g., load factor restrictions).

2.27 Limits have been imposed on the availability of each source of coal. For example, each deep-mined pit has a maximum production capacity set for each year and each port has a maximum coal handling capacity imposed on it.

Emissions Modelling

2.28 A useful by-product of developing the ESI model in this way is that sulphur emissions can be estimated on a plant-by-plant basis, since individual sulphur contents can be attached to the various coal types. Other types of emission can also, in principle, be estimated for each plant. This considerably improves the modelling of emissions and responses to various emissions abatement policies.

2.29 The modelling developments described here therefore improve estimates of electricity prices, individual plant behaviour, pool prices and not least, CO_2, SO_2 and nitrogen oxides (NOx).

Basic Modelling Procedure

2.30 The modelling changes described above do not alter in a fundamental way the basic approach to ESI modelling. The model is given a choice of plants, consisting of existing stations, those under construction and future possible types, from which it can choose to meet electricity demand. In some cases, plant categories include a number of individual plants. For example, existing oil plants are mostly aggregated into one plant group. The same applies to CCGTs, except that several types of CCGT are included in the list of plants, reflecting differences in performance (efficiency, etc).

2.31   For any given level of electricity demand, the model calculates the merit order of all available plants and runs them accordingly, subject to any constraints placed upon plant performance or emissions abatement either at individual plants or at a country or national level.

2.32   Given appropriate assumptions on future plant margins, the model needs to decide which types of new plant build will satisfy the growth in electricity demand. To do this, cost and performance data is required for each type of plant which can be built in the future. Once the level of electricity demand-mostly via the final demand equations - is estimated, the model then chooses how best to meet demand. A least-cost solution is generated, again subject to any specific constraints on plants, etc.

2.33   As with all models an ability to impose constraints reflecting the real world constrain is important. For example, the current coal contracts between the generators and the newly privatised coal companies are incorporated into the model. Apart from those constraints which reflect the real world, no further constraints are placed on the ESI model (e.g., future plant build).

Pool Prices

2.34   Given a merit order, pool prices are calculated, which, after the addition of various mark-ups to reflect transmission and distribution costs, the fossil fuel levy, value added tax (VAT) and normal profits, can be transformed into prices to final consumers for each final demand sector. These prices are then fed back into the energy demand models, which recalculate electricity demand and the process continues until a pre-specified degree of convergence is attained.

2.35   Having described the modelling process in general, the next chapter describes the assumptions used in the model.

CHAPTER 3. ASSUMPTIONS

3.1 In order to form a view of the possible levels of future energy demand, a number of assumptions need to be made. This section lists the main assumptions that underlie the energy demand scenarios and discusses the rationale for them. It also highlights differences in this area between EP65 and its predecessors, EP59 (published in 1992) and EP58[1] (published in 1990). The main assumptions about the ESI are reported in Chapter 5.

3.2 Like its two predecessors, EP65 uses ranges of assumptions, not single values. This is because of uncertainty about the future. Throughout EP65 a core set of six different scenarios has been used, augmented in some instances by two more extreme scenarios. The scenarios are chosen to encompass a plausible range of outcomes and assume that many features of the world remain unchanged (e.g. government policy). They are based purely upon different GDP growth and fuel price assumptions. These growth and price variants are designed to provide a feel for the range of likely outcomes and of the relative importance of growth and prices. No special significance is attached to any individual scenario. Table 3.1 below shows all eight scenarios and the abbreviations used to identify them in this paper.

Table 3.1 Scenario	Abbreviation
Central GDP Growth - High Fuel Prices	CH
Central GDP Growth - Low Fuel Prices	CL
High GDP Growth - High Fuel Prices	HH
High GDP Growth - Low Fuel Prices	HL
Low GDP Growth - High Fuel Prices	LH
Low GDP Growth - Low Fuel Prices	LL
Low GDP Growth - Very High Fuel Prices	LVH
High GDP Growth - Very Low Fuel Prices	HVL

3.3 The six main scenarios purposely take a fairly long run, smoothed view of fuel price developments. Because the actual range of uncertainty over shorter periods of time is rather larger than this, the scenarios with

---

[1] An evaluation of Energy Related Greenhouse Gas Emissions and Measures to Ameliorate them. Energy Paper 58, HMSO, 1990.

'very low' and 'very high' price variants aim to give an indication of the extreme range of energy demands in the period to 2000. Details of these extreme scenarios are not included in the main text, but results are shown in Annex G.

MAIN ASSUMPTIONS

Growth in Economic Activity

3.4   One of the most important determinants of energy demand in the UK has been the growth of economic activity. In these projections three different views of UK economic growth in the longer term have been examined: high, central and low growth. The average annual growth rates used in these scenarios are shown in Table 3.2. The table also shows the equivalent growth rates used in EP58 and EP59.

Table 3.2 Long Run GDP Growth Rate Assumptions			
	EP65	EP59	EP58
High Growth	2.85%	2.75%	3.25%
Central Growth	2.35%	2.25%	2.25%
Low Growth	1.75%	1.75%	1.25%

3.5   The growth rate in the central and high scenarios has been increased slightly in EP65, to make up for low growth in the early 1990s, so achieving the same level of GDP assumed for EP59 in the long run. A similar adjustment in the low GDP growth rate was unnecessary. The other features of the EP59 growth rate assumptions have been retained in EP65, with a 0.5% difference between the central and high growth rates. Moving the central and high rates up while keeping the same low rate has the effect of slightly increasing the gap between the central and low growth rates.

Sectoral Change

3.6   Different sectors or activities in the economy have different energy intensities. In order to produce scenarios of future energy requirements, assumptions are needed about the rate at which sectors or activities will grow or decline.

3.7   Until 2003 the assumed GDP structures have been based on scenarios developed by DTI using the Oxford Economic Forecasting industrial sector model. Such a model is however not suitable for very long-run projections, so beyond 2003 it has been assumed that most sectors' relative growth rates are similar to those experienced in the recent past. The main exception is the energy sector, where it is assumed that oil and gas production will decline significantly by 2020. Table 3.3 shows the assumed change in the structure of GDP in the central growth scenario over the projection period.

Table 3.3
Assumed change in GDP structure - Central GDP Growth Scenario[1]

	1990	1995	2000	2005	2020
Agriculture	21	21	20	20	20
Energy	73	86	88	85	62
Manufacturing[2]	409	391	404	406	392
Services	498	502	488	489	526
GDP	1000	1000	1000	1000	1000

1: Numbers represent each sector's weight per 1000 units of gross GDP.
2: Includes the Construction sector.

3.8   In EP59 no explicit allowance was made for the assumed GDP structure to vary between the different growth scenarios. While each sector's share of GDP changed over time, as it does in EP65, the share of GDP remained the same in the central, high and low growth scenarios. However, EP65 incorporates different GDP structures for each growth rate, as can be seen in Tables 3.4 and 3.5:

Table 3.4
Assumed change in GDP structure - High GDP Growth Scenario[1]

	1990	1995	2000	2005	2020
Agriculture	21	22	22	22	22
Energy	73	80	77	73	50
Manufacturing[2]	409	397	412	415	404
Services	498	501	489	490	524
GDP	1000	1000	1000	1000	1000

1: Numbers represent each sector's weight per 1000 units of gross GDP.
2: Includes the Construction sector.

Table 3.5
Assumed change in GDP structure - Low GDP Growth Scenario[1]

	1990	1995	2000	2005	2020
Agriculture	21	21	21	20	20
Energy	73	98	101	104	83
Manufacturing[2]	409	379	387	384	367
Services	498	502	491	492	531
GDP	1000	1000	1000	1000	1000

1: Numbers represent each sector's weight per 1000 units of gross GDP.
2: Includes the Construction sector.

3.9 The agriculture and service sectors' shares of GDP stay at similar levels in the central, high and low growth scenarios. Energy's share, on the other hand, varies considerably. When growth is low, energy's share of GDP is at its highest, and vice versa. This is because energy production from the North Sea, is not particularly sensitive to the rate of GDP growth in the UK.

3.10 The manufacturing sector's share of GDP also varies between the three different scenarios. However, in this case its share increases with the growth rate, i.e. its share is at its highest in the high growth scenario. This implies that it is the manufacturing sector which is responsible for much of the variation in the GDP growth rate.

3.11   The GDP structure used in EP59 followed the same basic trends over time as those in EP65, with the only real exceptions being in the energy and Manufacturing sectors. The long-run share of GDP attributable to energy is greater, and that attributable to manufacturing is lower, in EP65, largely as a result of an upward revision to future UK Continental Shelf oil and gas production.

Manufacturing Shares

3.12   Within the manufacturing sector, the energy intensity of different subsectors varies enormously. To take account of this it is necessary to make assumptions about changes in the structure of the manufacturing sector. These are shown in Table 3.6.

Table 3.6
Central GDP Growth Scenario - Manufacturing Shares[1]

Category	1990	1995	2000	2005	2020
Metals	36	34	35	34	27
Other Minerals	32	31	34	34	31
Chemicals	77	87	89	96	128
Food, etc.	139	147	133	130	136
Textiles & Clothing	39	37	37	35	28
Paper, etc.	71	77	72	72	76
Engineering	302	310	305	299	285
Other & Construction	302	278	294	300	288
Total Manufacturing	1000	1000	1000	1000	1000

1: Numbers represent each sector's weight per 1000 units of gross manufacturing output.

3.13 As with sectoral change, the structural change within the manufacturing sector is dependent upon the assumed rate of GDP Growth. Tables 3.7 and 3.8, below, show the structural change assumed within manufacturing shares in the high and low GDP growth scenarios.

Table 3.7
High GDP Growth Scenario - Manufacturing Shares[1]

Category	1990	1995	2000	2005	2020
Metals	36	36	35	33	25
Other Minerals	32	31	34	36	33
Chemicals	77	88	87	93	123
Food, etc.	139	148	135	132	138
Textiles & Clothing	39	38	36	33	28
Paper, etc.	71	76	71	69	73
Engineering	302	303	296	284	272
Other & Construction	302	280	305	319	308
Total Manufacturing	1000	1000	1000	1000	1000

1: Numbers represent each sector's weight per 1000 units of gross manufacturing output.

Table 3.8
Low GDP Growth Scenario - Manufacturing Shares[1]

Category	1990	1995	2000	2005	2020
Metals	36	33	34	34	29
Other Minerals	32	32	33	32	29
Chemicals	77	89	93	104	140
Food, etc.	139	150	145	140	147
Textiles & Clothing	39	38	36	34	26
Paper, etc.	71	79	77	78	83
Engineering	302	299	297	303	285
Other & Construction	302	280	285	274	261
Total Manufacturing	1000	1000	1000	1000	1000

1: Numbers represent each sector's weight per 1000 units of gross manufacturing output.

3.14 As with the GDP sectoral shares, the manufacturing shares in EP59 did not vary between growth scenarios. The definitions used for the various manufacturing sectors differ between EP65 and EP59. When allowance is made for these definitional differences similar basic trends are apparent in the EP65 and EP59 manufacturing shares.

3.15 The share of most manufacturing sectors across time is fairly constant or falling. The main exception is the chemicals sector, whose share of manufacturing increases between 1990 and 2020 in all three GDP scenarios in EP65. Recent historical experience suggests that this is not an unreasonable assumption: between 1982 and 1992 the chemicals sector's output increased on average by 4% per annum, easily outstripping even the service sector's output growth of 2.5% per annum[2]. After the chemicals sector, the paper, printing and publishing sector is assumed to have the next fastest rate of growth in output. Between 1982 and 1992 this sector grew at an average 3.3% per annum. The two other industry sectors with the slowest rate of increase in output during the projection period are textiles, leather and clothing and non-ferrous metals. All of the assumptions about the individual other industry sectoral growth rates are broadly reflected in the energy projections shown for each of these sectors in Annex A.

3.16 There is not a great difference in shares between the three growth scenarios for any of the sectors except the 'other and construction' sector, which displays a higher share in the high growth scenario.

Fuel Prices

3.17 There are four different energy price scenarios considered in these projections: very high, high, low and very low. The very high and very low energy price scenarios are designed to examine extreme energy price scenarios that could probably not be sustained in the long run. The crude oil price assumptions used in the scenarios are shown in Tables 3.9 and 3.10. For comparison purposes the equivalent values from EP59 and EP58 are included, where available, in the tables.

---

[2]Table 14.6, Annual Abstract of Statistics, Central Statistical Office, HMSO, 1994.

Table 3.9
Crude Oil Price Assumptions (1990 $ Barrel)

	EP65 H	EP65 L	EP59 H	EP59 L	EP58 H	EP58 L
1990	23.30	23.30	23.30	23.30	23.80	21.40
1995	20.00	15.00	27.00	17.00	39.30	22.60
2000	25.00	15.00	32.50	19.30	n/a	n/a
2005	27.50	15.00	35.00	21.20	n/a	n/a
2020	35.00	15.00	42.60	27.00	47.70	26.20

Table 3.10
Crude Oil Price Assumptions (1990 $ Barrel)

	EP65 Very High	EP65 Very Low
1990	23.30	23.30
1995	25.00	12.00
2000	40.00	10.00

3.18 The crude oil price assumptions used in the EP65 scenarios are lower than their EP59 and EP58 equivalents. Oil price expectations have been reduced in the light of experience since EP59. The change in the political environment in the last few years has helped to open up oil provinces to international investment from the oil companies. This has increased competition and sharpened the drive towards cost efficiency. These factors together with technological advances in exploration and development suggest that future oil demand can be met at a lower set of prices than was assumed in EP59.

3.19 World energy prices depend on the demand and supply for energy and in particular on the extent and sustainability of any market power held by energy producers. There is a great deal of uncertainty about these aspects of the world energy scene. Individual fuel producers' market power is constrained by the extent to which fuel substitution is possible when the relative prices of competitive fuels change. For crude oil the main competition comes from coal and gas. The price assumptions for coal and gas are shown in Tables 3.11 - 3.12 and 3.13 - 3.14 respectively.

Table 3.11
Steam Coal Price Assumptions (ARA 1990 $ tonne)

	EP65 H	EP65 L	EP59 H	EP59 L	EP58 H	EP58 L
1990	48.15	48.15	48.20	48.20	50.00	45.30
1995	45.00	40.00	50.00	45.00	n/a	n/a
2000	50.00	45.00	55.00	45.00	n/a	n/a
2005	51.15	45.60	55.00	45.00	64.30	52.40
2020	60.00	50.00	55.00	45.00	67.90	57.20

Table 3.12
Steam Coal Price Assumptions
(ARA 1990 $ tonne)

	EP65 Very High	EP65 Very Low
1990	48.15	48.15
1995	50.00	40.00
2000	60.00	40.00

3.20 In the short to medium term, the high coal price assumptions reflect an increase in shipping costs and enhanced world coal demand. Prices, by the end of the century, are projected to rise to a level which induces new entry in the US market. To meet forecast world demand over the long term could require construction of new mines in less favourable locations than current operations. An increasing number of mines with poorer geological features and situated further away from ports would result in higher costs and hence prices.

3.21 The low coal price assumptions depend on continued oversupply in the world coal market, particularly due the European recession, along with a rapid expansion of export capacity by low cost producers. Beyond the turn of the century new mining techniques are assumed to overcome the cost disadvantages associated with the less favourable sites. Delivered prices, by category, are based on the maintenance of recent historical marks up over the Amsterdam Rotterdam Antwerp (ARA) price.

Table 3.13
Gas Beach Price (1990 Pence per Therm)

	EP65 High	EP65 Low	EP59 High	EP59 Low
1990	17.3	17.3	17.3	17.3
1995	20.0	16.0	20.1	15.5
2000	26.0	20.0	20.9	16.7
2005	28.8	22.0	23.2	17.6
2020	37.0	25.0	31.4	22.8

Table 3.14
Gas Beach Price (1990 Pence per Therm)

	EP65 Very High	EP65 Very Low
1990	17.3	17.3
1995	26.0	16.0
2000	44.0	14.0

3.22 The gas beach prices used in these scenarios assume that the UK becomes fully integrated with the West European gas transportation system around the turn of the century and that until then prices are determined by UK costs. In the low price scenario prices are then assumed to trend up towards 25p/th in 2020, reflecting the price for marginal supplies. In the high price scenario, pressure from oil prices pushes gas prices up to 37p/th in 2020, around the level at which LNG becomes a viable alternative to gas.

Comparison with Other Sources

3.23 Tables 3.15 shows how the EP65 crude oil price assumptions compare with those used in some other projections.

Table 3.15				
Crude Oil Prices (1990 $ Barrel)				
	1990	1995	2000	2005
EP65 (High)	23.3	20.0	25.0	27.5
EP65 (Low)	23.3	15.0	15.0	15.0
IEA[1]	22.0	19.9	24.8	27.2
EC[2]	21.6	19.0	23.0	28.0
SEEC[3]	23.3	14 - 15	14 - 15	14 - 15
CE[4]	23.7	14.5	17.8	20.2

1. International Energy Agency - "World Energy Outlook" (Paris, 1993)
2. European Community - "Energy in Europe - A View to the Future" (Brussels, Sept. 1992)
3. Surrey Energy Economics Centre - "The SEEC UK. Energy Demand Forecast (1993-2000)" (University of Surrey, December 1993)
4. Cambridge Econometrics - "Prospects for Reducing CO2 Emissions to the year 2020 - A report submitted to the National Steering Committee of the Nuclear Free Local Authorities" (Cambridge, August 1994)
5. Different definitions of oil price are used by each forecasting group.

3.24 The crude oil price assumption used by the International Energy Agency (IEA) and those used by the European Community (EC) are both similar to the high oil price assumptions used in EP65. The prices in these sources all lie within the EP65 price range except for the EC's price in 2005. The crude oil price used by the Surrey Energy Economics Centre, on the other hand, is almost identical to the low oil price series used in EP65. The price used by Cambridge Econometrics lies within the EP65 range for most years, but nearer the low price series.

Table 3.16
Steam Coal Prices ARA[1] (1990 $ tonne)

	1990	1995	2000	2005
EP65 (High)	48.15	45.00	50.00	51.15
EP65 (Low)	48.15	40.00	45.00	45.60
IEA[2]	50.90	45.40	48.30	51.30
EC[3]	54.30	50.00	52.00	55.00
CE[4]	68.00	51.20	48.40	45.00

1. Different definitions of coal prices are used by each forecasting group.
2. International Energy Agency - "World Energy Outlook" (Paris, 1993)
3. European Community - "Energy in Europe - A View to the Future" (Brussels, Sept. 1992)
4. Cambridge Econometrics - "Prospects for Reducing CO2 Emissions to the year 2020 - A report submitted to the National Steering Committee of the Nuclear Free Local Authorities" (Cambridge, August 1994)

3.25 Table 3.16 compares the EP65 steam coal price with three other sources. The IEA's prices are again very similar to the EP65 high prices, with the EC prices slightly higher than the EP65 high price scenario for all the years examined. The CE coal price series differs from that used in EP65, being higher than the EP65 high price case in the early years and right at the bottom of the EP65 range in the later years.

Prices to Consumers

3.26 This section looks at the assumptions behind the prices to consumers of the different types of oil, solid fuel and gas.

3.27 In broad terms energy prices to final users are either flat or falling in the low price scenarios and rising only modestly in the high price scenarios. In the extreme scenarios (very high and very low prices) the prices rise/fall steeply over a short period. It should be noted that these two scenarios are used only up to the year 2000 in the current projections. The long-term projections use only the high and low energy price scenarios.

Final User Oil Price Assumptions

3.28 Sterling oil product prices are chiefly determined by the crude oil price and exchange rate assumptions. UK tax exclusive product prices have been determined by applying ratios to the sterling crude oil price. The ratio for each product is determined judgmentally, taking into account historic trends and likely changes in product markets. To derive consumer prices, assumptions are required for distribution costs and taxes. Distribution costs are assumed to stay constant in real terms throughout the period. The oil product tax structure is assumed to remain unchanged, except in the domestic sector and the road transport sector. The decision of 6 December 1994 to proceed with VAT on domestic fuel at 8% has been accounted for, as have the changes in the duties on road fuels, heavy fuel oil and gas oil announced on 9 December 1994. The 5% per annum real increase in road fuel duties has, for the purposes of these projections been assumed to apply until the year 2000 inclusive.

3.29 Nine final user oil prices are identified in the energy model:

1. Diesel Engine Road Vehicle Fuel (DERV) price
2. Motor Spirit price
3. Aviation Fuel price
4. Water Transport Gasoil price
5. Other Industries Heavy Fuel Oil price
6. Iron & Steel Industry Heavy Fuel Oil price
7. Domestic Sector Average price of Petroleum[3]
8. Service Sector Average price of Petroleum[1]
9. Agriculture Sector Average price of Petroleum[1]

3.30 Chart 3.1 shows the road transport final user fuel prices in the high and low energy price scenarios. Chart 3.2 does the same for heavy fuel oil (applicable to all industrial sectors, including iron & steel). Finally, Chart 3.3 shows the average price of petroleum (which in this context means all oil based fuels) in the domestic and services sectors in the same two oil price scenarios.

---

[3] "Petroleum" includes all oil products except for domestic and service sector use of derv and motor spirits (included in the transport sector).

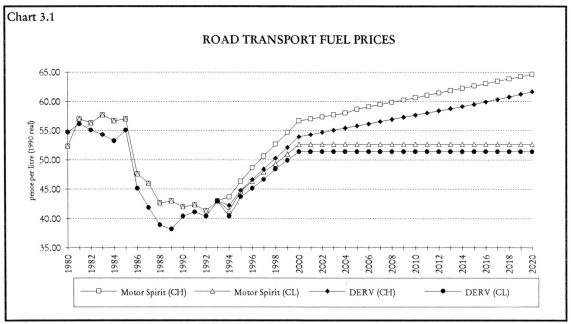

Note: the Motor Spirit price used before 1990 is the 4-star price. After 1990 a weighted index of 4-star and unleaded prices is used.

3.31   In the low oil price scenarios the 5% per annum real increase in road fuel duties can be seen to dominate the movement in road transport's final prices. Beyond the year 2000 the flat $15 per barrel price assumption produces flat final road transport prices. In the high oil price scenarios the increase in road fuel duties accompanies a rise in the price of crude oil throughout the projection period.

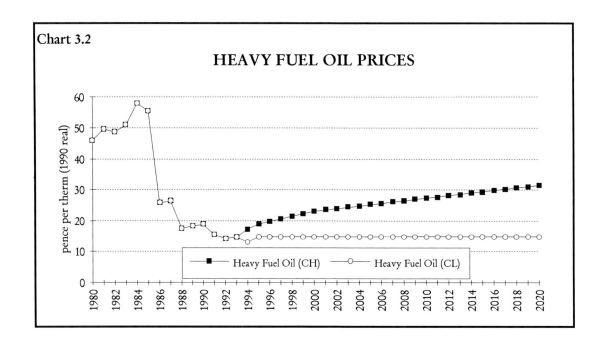

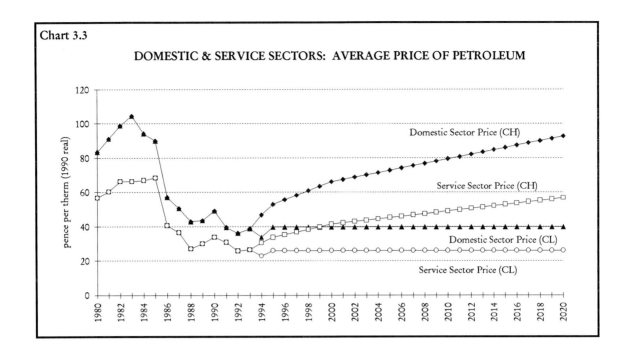

Chart 3.3
DOMESTIC & SERVICE SECTORS: AVERAGE PRICE OF PETROLEUM

Final User Gas Price Assumptions

3.32 The marginal beach and the weighted average cost of gas (WACOG) prices have been used, along with information taken from the recent Monopolies and Mergers Commission (MMC) report on British Gas[4], to produce the final user gas price assumptions. The marginal beach price is the incremental price of new gas supplies. The nominal WACOG is an average price of all gas supplies, with relevant weights allocated to the different sources according to how much gas is obtained from each source.

3.33 Currently, there are three different types of gas contract available to final users: interruptible, firm and tariff contracts. Interruptible gas is mainly used in the industrial sector, but also to a lesser extent in the service sector. Firm gas is used in the industrial and service sectors. The tariff market, is dominated by the domestic sector.

3.34 An average gas price for each of the three sectors has been obtained by allocating weights to the different types of gas used in that sector. For example, an industrial sector average gas price has been calculated by weighting the tariff, firm and interruptible contract prices appropriately.

---

[4]Gas and British Gas plc - Volume 2 of reports under the Gas and Fair Trading Acts, Monopolies and Mergers Commission, HMSO, September 1993.

3.35   The projected service sector average gas prices have been calculated in a similar manner to the industrial sector average gas prices, except that higher weights have been applied to the firm and tariff contract prices than in the industrial sector.

3.36   The domestic prices used in the energy model are for the 800 therm and 400 therm consumption bands. Historic data can be found in table 58 of the Digest of UK Energy Statistics. For projection purposes the 1993 800 and 400 therm gas prices have been indexed with the tariff prices taken from the OFGAS regulatory formula[5]. The real domestic gas prices include the imposition of VAT at 8%.

3.37   Charts 3.4, 3.5 and 3.6 show the assumed average final use gas prices for the domestic (800 therms), service and industrial sectors respectively.

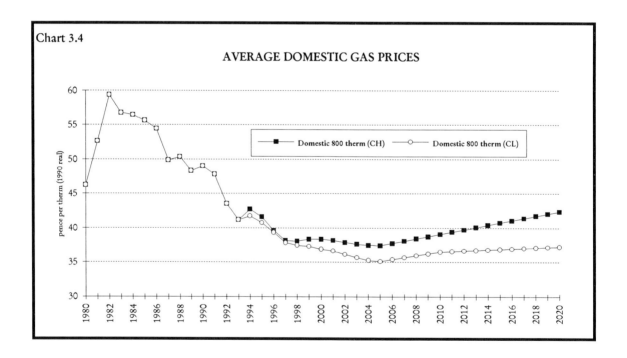

3.38   The domestic gas prices shown in Chart 3.4 are tariff prices. Up to 1998 these are controlled by regulatory formula. The prices are essentially composed of two elements; a non-gas element and a gas element. The non-gas element of the tariff price is of the form Retail Price Index (RPI) - X. By

---

[5]See MMC op cit., pages 106-110

contrast, the gas element is determined by a gas cost index (GCI), which is a basket of prices designed to reflect contract escalation arrangements. The GCI is composed of several elements such as exchange rates, inflation, a crude oil price, heavy fuel oil prices and gasoil prices. The 1998 gas element of the tariff price is calculated from the GCI, minus 1% per annum to reflect existing regulatory controls. Beyond 1998 the gas element of the tariff price is indexed with the weighted average cost of gas.

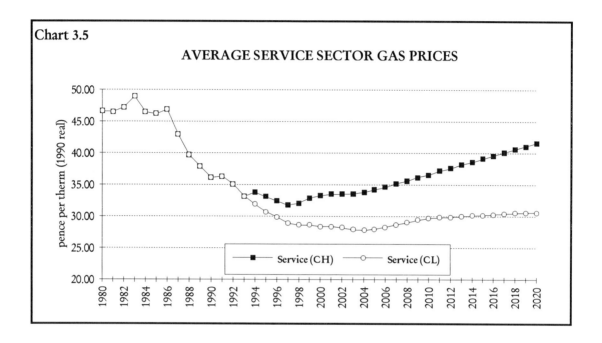

Chart 3.5
AVERAGE SERVICE SECTOR GAS PRICES

3.39 The service sector gas prices shown in Chart 3.5 are a weighted average of firm, tariff and interruptible gas prices. Initial projections of industrial firm and interruptible gas prices were created via indexation with a WACOG. Service sector projections of firm and interruptible gas prices were then computed by applying the ratio of commercial to industrial gas prices in 1992 shown in table 3.5 of volume 2 of the MMC report.[6]

---

[6]See MMC, op. cit., table 3.5 page 69.

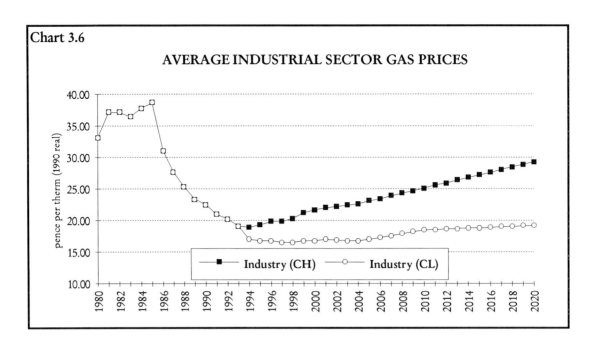

Chart 3.6 — AVERAGE INDUSTRIAL SECTOR GAS PRICES

3.40 The industrial gas prices shown in Chart 3.6 are a weighted average of interruptible, firm and tariff prices. The industrial interruptible and firm gas price projections were created via indexation with the WACOG. It can be seen from Chart 3.6 that the average industrial gas prices produced by this method, for the period 1993 - 2020, are either flat in the low energy price scenarios or grow at 1.6% per annum in the high energy price scenarios.

Final User Solid Fuel Price Assumptions

3.41 The projections for the final use price of solid fuels are again split into different end-use categories. The four different prices used are:

1. Other industry coal price
2. Iron and steel industry coal price
3. Domestic sector solid fuel price
4. Service sector solid fuel price

3.42 The coal price paid by the iron and steel industry is assumed to be the same as that paid by other industries. This assumption reflects the availability of historic data rather than a detailed analysis of the coal prices paid in both sectors. The service sector, because of its larger average consumption, pays a lower solid fuel price per therm than the domestic sector.

Assumptions

3.43 The projections shown include the 5% per annum real increase in road fuel duties and all of the road fuel duty increases announced in November and December 1994. The car stock assumptions behind the road transport projections are consistent with the 1989 National Road Traffic Forecast produced by the Department of Transport.[7] A maximum penetration for diesel powered cars, out of the total car stock, of 25% has been imposed on the road transport model.

Household Assumptions

3.44 Chart 3.7 shows the assumed number of UK households over the projection period. This number has been derived from population projections in the 1993 Annual Abstract of Statistics table 2.5 and a DTI projection of average household size. This projection of average household size was obtained by indexing 1990 mean household size data from the General Household Survey with the BRE's average household size projection.

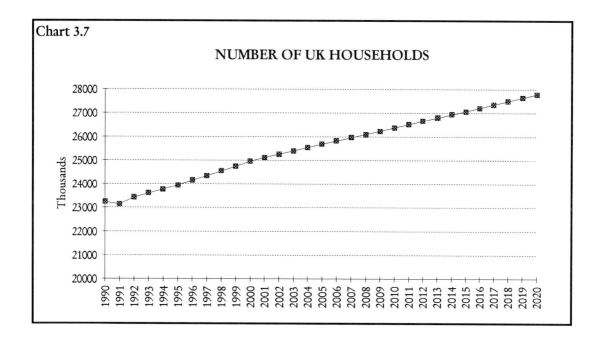

---

[7]National Road Traffic Forecast, Department of Transport, HMSO, 1989.

Exchange Rates

3.45 The oil and coal international price assumptions used in the projections are expressed in US dollars to reflect the dollar's widespread use as a numeraire in international energy spot markets. Since UK consumers pay for their energy in sterling it has been necessary to make assumptions about the real sterling/dollar exchange rate. These assumptions are shown below:

Table 3.17	
Year	Real $/£ Exchange Rate
1990 (base)	1.79
1991	1.8
1992	1.8
1993	1.52
1994 - 2040	1.52

3.46 The long-run $/£ exchange rate of 1.52 compares with a figure of 1.65 for EP59. The lower sterling/dollar exchange rate implies that, for any given set of US dollar energy prices, UK sterling energy prices will be higher.

CO_2 EMISSIONS ASSUMPTIONS

3.47 The Government, as part of its 1994 CCP, has introduced a series of measures aimed at enabling it to return CO_2 emissions to 1990 levels in the year 2000. On the basis of the EP59 projections, the working assumption was made that this would involve a reduction in CO_2 emissions of around 10 MtC. When it was published in January 1994 the CCP estimated that the various measures would give rise to carbon savings of this magnitude.

3.48 The CCP measures, amended for the tax changes announced in November and December 1994, have been included in the projections: the tax measures have been directly incorporated into the fuel price assumptions (8% VAT on domestic energy and 5% average real increase in road fuel duties each year); for the other measures, the relevant *energy* savings assumed in the CCP have been imposed on the model, rather than the CO_2 savings themselves.

3.49 This means that in the projections the carbon savings attributable to the CCP are 1 - 2 MtC lower than those originally anticipated when the CCP was published (shown in Table 3.18). This is because some of the measures have been changed (e.g. VAT on domestic energy at 8%) and because the properties of the current projections models (fuel mix, underlying fuel prices, price elasticities etc.) are not identical to those assumed when the impact of the measures was initially calculated. The Department of the Environment is now reviewing the implications of the projections for the achievement of the UK's CO_2 target in the year 2000.

Table 3.18
Carbon Savings Originally Assumed in the Climate Change Plan

Sector	Reduction in emissions by 2000, MtC
Energy consumption in the home	4
Energy consumption by business	2.5
Energy consumption in the public sector	1
Transport	2.5
Total	10

Note: The EST was assumed to account for 2.5 MtC of the total 10 MtC; VAT at 17½% on domestic fuel and power was assumed to account for 1.5 MtC.

CHAPTER 4. FINAL USER DEMAND PROJECTIONS

4.1 This chapter describes the main features of the six long-run final user energy demand projections. Final user demand excludes energy use and losses in the energy industries themselves (such as conversion losses in electricity generation). Primary demand includes them and is dealt with in Chapter 6. The detailed final user projections can be found in Annex A. Projections of the domestic and service sector broken down by end-use are shown in Annex B (heating, cooking and appliances etc)[1].

DOMESTIC SECTOR ENERGY PROJECTIONS

Table 4.1	Domestic Sector				Billions of Therms			
	1990	1991	1995	2000	2005	2010	2015	2020
LL	16.2	18.0	18.0	18.0	18.3	18.4	18.5	18.7
LH	16.2	18.0	17.8	17.7	17.9	17.9	18.0	18.1
CL	16.2	18.0	18.0	18.1	18.4	18.5	18.6	18.8
CH	16.2	18.0	17.8	17.7	18.0	18.0	18.1	18.2
HL	16.2	18.0	18.0	18.1	18.4	18.5	18.7	18.8
HH	16.2	18.0	17.8	17.7	18.0	18.0	18.2	18.3

4.2 Because the Government's carbon dioxide programme aims to return carbon dioxide emissions to their 1990 level by 2000, the starting point for the final user energy demand projections is 1990. However, a very warm winter in 1990 resulted in an unusually low level of domestic sector energy demand. For this reason a more typical year against which to compare the projections is 1991, when a more typical winter occurred. Note that a long-run average temperature has been assumed for all projection years post-1993.

4.3 It can be seen from Table 4.1 that domestic sector energy demand is projected to remain broadly unchanged from the 1991 level. These projections incorporate the impact of the 1994 CCP measures, such as VAT at 8%, new building regulations and the Energy Savings Trust schemes. In

---

[1] In common with the IEA and the Statistical Office of the European Communities, the Digest of UK Energy Statistics has recently switched from therms to tonnes of oil equivalent (toe) as the common energy unit of measurement. To convert between these two the conversion factor is 1 tonne of oil equivalent = 397 therms.

addition only 90% of UK households are assumed ever to have access to the gas distribution network and this effectively limits the long run growth of future gas demand. Consumption of electricity via the major appliances, such as washing machines and fridges, is also limited by the constraint that ownership levels of these appliances cannot exceed 100% of households. These and other similar 'saturation' effects largely explain why the effect on energy demand of steadily rising consumer income tends to diminish in the long run, so that the domestic sector's energy projections are more sensitive to price than to income.

Heating Related Energy Demand

4.4   The underlying growth in heating related energy demand reflects the tendency of households to increase the internal temperature of their dwellings as incomes rise. Households may elect, for instance, to heat all rooms up to the same standard previously reserved for the main living room.

Solid Fuels
4.5   Within the aggregate domestic sector projections solid fuel demand continues to decline from its 1991 level of 2,110 million therms (8.11 million tonnes). By the year 2000 domestic sector solid fuel demand is projected to be a little over half of the 1991 level. Solid fuel demand continues to decline until the end of the projection period in 2020 when it is around 14 - 21% of its 1991 level.

Petroleum Products
4.6   Petroleum product demand is similarly projected to decline in future, albeit with some minor short term increase following the recent increase in the number of households with oil fired heating. By the year 2000 domestic sector petroleum demand is projected to decline to some 34 - 53% of its 1991 level of 1,090 million therms (2.46 million tonnes). Domestic sector petroleum demand then continues to fall beyond the year 2000 to insignificant levels in all scenarios.

Gas and Electricity
4.7   Gas is the principal beneficiary of the decline in petroleum products and solid fuel demands. Electricity also benefits from the decline in

petroleum products and solid fuel demands, particularly in the non-gas areas of the UK, such as rural areas and Northern Ireland. Electricity is also increasingly used in small (particularly single-person) households where it is most competitive with gas. Electricity demand for heating purposes therefore grows at a modest rate in all scenarios because of the trend towards smaller households and fuel switching.

Appliance Energy Demand

4.8   Appliance related electricity demand remains broadly flat at its 1991 level until 2005 across all scenarios as higher appliance ownership is offset by improvements in appliance efficiency resulting in part from the 1994 CCP measures. Beyond 2005 further improvements in the efficiency of the existing stock of appliances becomes more limited and the introduction of new appliances leads to modest long-term growth in demand. New and minor appliances are not explicitly identified in the model; instead energy demand associated with these appliances has been projected forward on the basis of the historic relationship between demand and real personal disposable income. Appliances not currently widely used in the home, such as personal computers and fax machines, fall within this category.

4.9   Although all appliances are assumed to run on electricity it is reasonable to suppose that gas could penetrate the domestic appliance market. An example of this is gas tumble driers, which have already made inroads into the domestic clothes drying market. Further penetration of the domestic drying market could reasonably be expected to reduce UK CO_2 emissions. The assumption that all of the above appliances continue to run on electricity is therefore a cautious one.

Cooking-Related Energy Demand

4.10   Energy for cooking purposes (not microwaves, which are dealt with under appliances) is expected to decline in total until 2010. This decline in cooking demand reflects growing use of microwaves, improved cooker efficiency and more eating out. (The continuing trend towards greater eating out is also reflected in the service sector projections, which show an offsetting increase in cooking-related energy demand.) Beyond 2010 the scope for further improvements in cooker efficiency diminishes and is eventually outweighed by the effect of the growth in household numbers. Throughout the projection period gas increases its share of total cooking energy demand at the expense of electricity. Cooking is the only end use for which electricity demand declines during the projection period.

Comparison with Energy Paper 59

4.11   In both these projections and those reported in EP59 the ownership of central heating and major appliances in the domestic sector approaches saturation levels in the long run. Both sets of projections show relatively flat domestic sector energy demand into the future. For the year 2000 EP59 projected that domestic sector energy demand would be around 17 billion therms in all scenarios, compared to a range of 17.7 - 18.0 billion therms reported here. This slight increase can be attributed mainly to better modelling of the effects of temperature. When the EP59 model was being developed in early 1991, the 1990 domestic sector energy demand data was still very provisional and subsequent analysis of energy demand during the 1990 and 1991 winters has suggested a bigger temperature effect than was previously assumed. The increase in projected energy demand resulting from better temperature modelling has however been partly offset by the inclusion of the 1994 CCP measures.

4.12   For the year 2020 the EP59 projections suggested a range of 17 - 18 billion therms, compared to 18.1 - 18.8 billion therms in Table 4.1. The new domestic sector energy demand projections are therefore just above the top end of the range indicated in EP59. Again the improvement in temperature modelling explains most of the difference.

CHART 4.1

EP65 DOMESTIC SECTOR ENERGY DEMAND PROJECTIONS

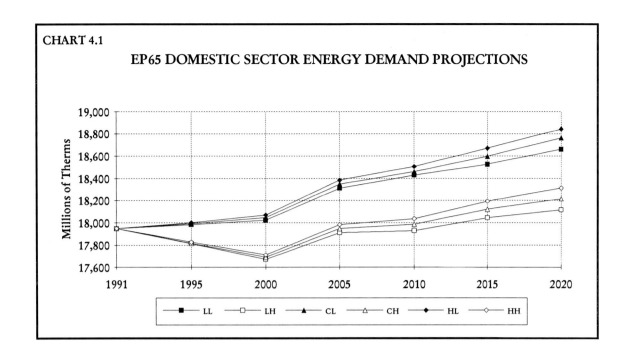

4.13 The impact of the 1994 CCP measures is evident from Chart 4.1 as their gradual introduction over the period to 2000 partly offsets the underlying growth in domestic sector energy demand. Once the 1994 CCP measures achieve their full impact (assumed to be in 2000) the associated reduction in energy demand remains broadly constant. Thus beyond 2000 the underlying growth in domestic sector energy demand reasserts itself.

IRON AND STEEL SECTOR ENERGY PROJECTIONS

Table 4.2		Iron & Steel Sector			Billions of Therms		
	1990	1995	2000	2005	2010	2015	2020
LL	3.0	2.7	2.9	3.2	3.1	3.1	3.0
LH	3.0	2.7	2.9	3.2	3.2	3.1	3.0
CL	3.0	2.8	3.3	3.6	3.7	3.7	3.6
CH	3.0	2.8	3.3	3.7	3.7	3.7	3.5
HL	3.0	3.0	3.4	3.7	3.8	3.8	3.7
HH	3.0	2.9	3.4	3.7	3.8	3.8	3.7

4.14 The iron and steel sector's energy demand is dominated by the quantity of crude steel produced (17.8 million tonnes in 1990). Crude steel is currently produced in the UK using two main types of furnace, the basic oxygen steel (BOS) furnace and the electric arc furnace (EAF); the BOS furnace's main energy input is coke, the EAF's is electricity. In 1990 approximately 75% of crude steel production came from BOS furnaces and the remaining 25% from EAFs.

4.15 One of the most notable factors about the projected iron and steel energy demands shown in Table 4.2 is the limited impact of prices on aggregate energy demand. Assumptions about the future level of UK crude steel production are the main determining factor behind the EP65 iron and steel sector's energy demand projections.

4.16 Chart 4.2 emphasises the relatively flat nature of future energy demand in the iron and steel sector. Some recovery is assumed from the current low levels of crude steel production with the results that by the year 2000 energy demand is expected to be higher than in 1995. Compared to 1990 the range for 2000 is -3% to +13%. Since the bottom of the projected energy demand range is lower than the 1990 level the projections implicitly assume that improvements in energy efficiency will continue during this period.

4.17 By 2020 the EP65 range is up to 25% greater than the 1990 figure. Five of the six scenarios suggest that energy demand in 2020 will be lower in

2005. Since crude steel production is assumed not to decline between 2005 and 2020 in any of the scenarios there is an implied continuation of trend improvements in efficiency.

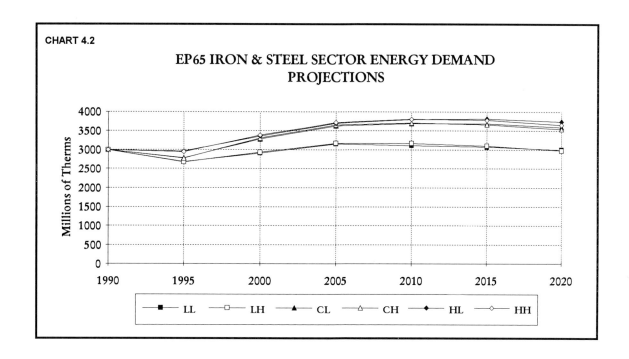

OTHER INDUSTRY ENERGY PROJECTIONS

Table 4.3		Other Industry Sector		Billions of Therms		
	1990	1995	2000	2005	2010	2015	2020
LL	12.2	11.7	12.4	12.9	13.1	13.3	13.7
LH	12.2	11.5	11.6	12.1	12.2	13.0	14.0
CL	12.2	11.8	12.9	13.8	14.0	14.6	15.4
CH	12.2	11.6	12.1	12.9	13.2	14.4	15.9
HL	12.2	11.9	13.2	14.3	14.7	15.5	16.6
HH	12.2	11.7	12.4	13.4	13.9	15.4	17.2

4.18   Chapter 2 has described how the other industry sector is composed of eight separately modelled sectors - namely non-ferrous metals, mineral products, chemicals, engineering & vehicles, food drink & tobacco, textiles leather & clothing, paper printing & publishing and construction & other industries. Because there has always existed some room in the statistics for misallocation of individual fuels between the eight sectors, only total projected energy demand is shown here for each of the sectors. However, projections of coal, oil, gas and electricity are shown for other industry (i.e. the sum of the eight industrial sectors) in Annex A.

4.19   Table 4.3 indicates that other industry's projected energy demand in the year 2000 has a range of -5% to +9% when compared with 1990. Both other industry's output and energy prices can be seen to have an impact on the projected energy demands for the year 2000. Unlike in the iron and steel sector it is the variation in energy prices, rather than in output, which produces the largest difference between the scenarios. With the exception of non-ferrous metals in the low GDP scenarios, output in each of the eight sub-sectors is assumed to be higher in 2000 than in 1990.

4.20   By 2020 other industry's energy demand is projected to be 13% to 41% higher than in 1990. This implies that during the period 1990-2020 other industry's energy demand grows by 0.4% to 1.2% per annum. An interesting aspect of the 2020 projections is that energy demand is higher in the high energy price scenarios rather than in the low energy price scenarios. This higher demand in the high energy price scenarios arises because it is *useful* energy demand that is actually modelled and not the *delivered* energy demands as reported in Table 4.3.

4.21   Although the *absolute* increase in all energy prices reduces total useful energy demand the *relative* energy prices in the high energy price scenarios strongly favour coal. Since coal has the lowest conversion efficiency of the four basic types of energy, a large increase in the useful therm share of coal can, if demand is sufficiently sensitive to relative fuel prices, in principle result in an absolute increase in total delivered energy demand when the useful therm coal demand is converted into delivered therms. Thus it is the switch back into coal in the high energy price scenarios that results in projected delivered energy demand being greater than in the low energy price scenarios at the very end of the projections period.

4.22   In view of the difficulty of accurately measuring average conversion efficiencies across a wide range of end uses and of identifying cross-price elasticities so far into the future, a greater than usual degree of uncertainty must be attached to this result. If the conversion efficiencies of different fuels turn out to be closer to each other than has been assumed or if the responsiveness of demand to relative prices is less, then this would reduce the total delivered energy demand projections in the high price cases, with consequences for the CO_2 projections reported in Chapter 7.

Non-Energy Related Fossil Fuel Demand

4.23   Both gas and oil are used in industry for purposes other than to produce heat or motive power. For example, gas is used to make fertiliser and oil is used as a feedstock. From 1995 - 2020 it is assumed that non-energy demand for gas is 785 million therms per annum and demand for non-energy oil is 10 million tonnes of oil per annum The use of gas and oil for these purposes is usually thought of as being related to industrial processes industry. However, following the definition used in DUKES, non-energy data is not included in the other industry projections shown above. The non-energy related fossil fuel projections can however be found in the total final user tables of Annex A.

Comparison with Energy Paper 59

4.24 A direct comparison between EP59 and the projections reported here is not easy as the fossil fuels used to produce own generated electricity are no longer shown in the final user tables of DUKES. Instead of recording the fossil fuels used to produce the own generated electricity, the statistics now record the electricity produced by the own generation plant. This convention has also been adopted here. In addition EP59 only reported energy demand projections for industry as a whole and, since this is the sum of the other industry and iron and steel sectors' demands, these sectors' have been amalgamated in Table 4.4 to enable a comparison to be made.[2]

Table 4.4	Industry EP65 - EP59		Billions of Therms	
	1995	2000	2005	2020
LL	-3	-3	-2	-1
LH	-2	-1	0	2
CL	-3	-2	-1	0
CH	-3	-1	1	3
HL	-3	-2	-2	-1
HH	-2	-1	0	2

Note: the EP65 and EP59 projections treat fossil fuels used in autogeneration differently. Putting the EP65 projections on the same basis as those in EP59 would increase the former.

4.25 Table 4.4 shows the new industrial sector projections minus the equivalent EP59 projection for each scenario. In the context of total industrial energy demand, these differences are small: the new industrial sector projections are of the same broad magnitude in the long-run as their EP59 counterparts. Data definitional differences aside, it can be seen that in the short run the new projections are slightly lower than the equivalent EP59 projection mainly because of lower industrial output. In the low price scenarios, this difference persists, albeit diminishingly so, into the long run. In the high price scenarios, the improvement in the competitiveness of coal is associated with a stronger coal demand and hence a higher level of delivered energy demand than in EP59. The caveats made above about the robustness of this result apply.

---

[2]The EP59 (unlike the EP65) industrial sector projections included non-energy use of gas.

SERVICE SECTOR ENERGY PROJECTIONS

Table 4.5		Service Sector			Billions of Therms		
	1990	1995	2000	2005	2010	2015	2020
LL	7.8	8.3	8.7	9.1	9.4	9.8	10.2
LH	7.8	8.3	8.6	8.9	9.3	9.7	10.0
CL	7.8	8.4	8.9	9.3	9.8	10.3	10.8
CH	7.8	8.3	8.7	9.2	9.7	10.2	10.6
HL	7.8	8.4	9.0	9.5	10.1	10.7	11.3
HH	7.8	8.3	8.8	9.4	10.0	10.6	11.1

Note: the EP65 and EP59 projections treat fossil fuels used in autogeneration differently. Putting the EP65 projections on the same basis as those in EP59 would increase the former.

4.26 The service sector projections shown in Table 4.5 and Annex A cover the public administration (schools, hospitals, local government etc), commercial and agricultural sectors. Detailed end-use projections for the first two sub-sectors are shown in Annex B. The agricultural sector's energy demand is 500 - 600 million therms per annum in every scenario.

4.27 Table 4.5 indicates that total service (including agriculture) sector energy demand in the year 2000 is projected to be 10% to 15% higher than in 1990. By 2020 the range of projections suggest that service sector energy demand may be 28% to 45% higher than in 1990. Within this total, electricity and gas generally increase their share at the expense of solid fuels and petroleum products. To understand the projected growth in service sector energy demand each of the five main end-uses is examined below.

Cooking-Related Energy Demand

4.28 As described in the domestic sector section of this chapter, the trend towards eating out and towards the consumption of prepared foods is assumed to continue throughout the period 1990 - 2020. This implies that the energy demand for cooking in the service sector should be increasing during the projection period. Annex B shows the detailed projections for cooking-related energy demand. Both cooking fuels, gas and electricity, are projected to increase during the projection period. Electricity is expected to grow slightly faster than gas and becomes the dominant service sector cooking fuel early in the next century.

Lighting-Related Energy Demand

4.29   Electricity demand for lighting purposes (retail display, office lighting etc,) is projected to increase by 1.1% to 1.4% per annum throughout the period 1990 - 2020. Service sector output growth is the main factor determining this growth in lighting electricity demand, as both own-price and inter-fuel substitution effects have historically exerted little influence on demand.

Air Conditioning-Related Energy Demand

4.30   Electricity demand for air conditioning increases by 2.9% to 3.3% per annum throughout the period 1990 - 2020. In recent years electricity has met all the demand for air conditioning and it is assumed in the projections that it will continue to do so. In the past gas has been used for air conditioning and there is some evidence to suggest that gas may re-enter this market in the UK, as it has already done in the US. Since gas is a relatively clean fuel this would almost certainly have a beneficial impact on UK emissions of CO_2. The assumption that air conditioning energy demand is met by electricity rather than gas is therefore a cautious one.

Other Appliance Related Energy Demand

4.31   Electricity demand for other uses (photocopiers, personal computers, printers and fax machines etc) has increased dramatically during recent years. During the projection period this growth is unlikely to be sustained as more efficient appliances and saturation effects begin to limit the growth in electricity demand associated with these appliances. In the US, for instance, the Energy Star programme is already reducing the power consumption of printers and personal computers. A by-product of the US Energy Star programme is that these energy efficient appliances have started to become available in the UK, thereby also reducing electricity demand in this country. Although the service sector model explicitly allows for this effect, other electricity demand is still projected to grow at 0.8 - 1.5% per annum during the period 1990 - 2020. In the short-run (1990 - 2000) the projected growth is much higher at 3.6 - 4.6% per annum.

Heating

4.32   Heating is the only end-use in the service sector still to use solid fuels and petroleum products. Both of these fuels have however been declining in importance for a number of years and this trend is expected to continue in all scenarios. Gas seems likely to increase its dominance and is projected to approximately double by 2020. Although electricity for heating purposes is also expected to increase during the period 1990 - 2020, a more modest 25% growth is expected.

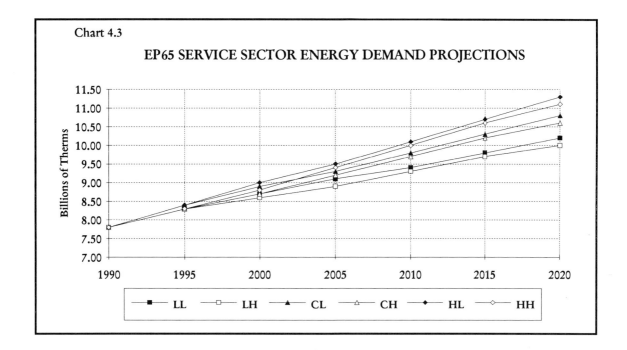

4.33   Chart 4.3 shows total service sector energy demand increasing from a little under 8 billion therms in 1990 to around 10.0 - 11.1 billion therms in 2020, or 0.8 - 1.4% per annum.

Comparison with Energy Paper 59

4.34   The projections for 2000 of 8.6 - 9.0 billion therms are at the bottom of the EP59 range of 9 - 10 billion therms. Given that the new projections include both the 1994 CCP measures and lower output in the 1990s this result was to be expected.

4.35   A comparison of the EP65 projections for the year 2020 reveals larger differences. At 10.0 - 11.3 billion therms, the new range is significantly lower than the EP59 range of 12 - 21 billion therms.

4.36   This difference of up to 10 billion therms is largely attributable to the more disaggregated approach now being used, compared to the more aggregated econometric approach adopted previously. The reduction in the year 2020 is overwhelmingly electricity. Incorporating stock and saturation effects in the other and air conditioning electricity demand models has weakened the link that existed in EP59 between service sector output and electricity demand. This large reduction in electricity demand in 2020 explains a significant part of the difference in CO_2 emissions between EP65 and EP59 shown in Annex E.

TRANSPORT SECTOR ENERGY PROJECTIONS

Table 4.6		Transport Sector			Billions of Therms		
	1990	1995	2000	2005	2010	2015	2020
LL	19.3	20.9	22.3	23.8	25.8	28.5	31.6
LH	19.3	20.7	21.7	22.6	24.0	26.2	28.8
CL	19.3	21.0	22.9	24.9	27.4	30.7	34.6
CH	19.3	20.8	22.2	23.7	25.6	28.4	31.7
HL	19.3	21.1	23.4	26.1	29.1	33.0	37.5
HH	19.3	20.9	22.7	24.9	27.3	30.7	34.6

Projections

4.37 The transport sector's energy demand, shown in Table 4.6, is projected to grow by 1.2% to 1.9% per annum for the period 1990 - 2000. In the longer term, 1990 - 2020, growth increases slightly to give a range of 1.3% to 2.2% per annum. Greater car ownership and increased air travel are the main factors behind this increase in the transport sector's energy demand.

4.38 These projections need to be interpreted with some caution: the estimates of future aviation fuel demand imply that growth continues into the long run at rates similar to those experienced in recent years. It is possible that such growth rates are unsustainable, in which case the demand for fuel from the transport sector at the latter end of the period would be less than shown here (this would not greatly affect the CO_2 projections, as most of the emissions associated with air transport are excluded from the current definition of UK CO_2 emissions).

Comparison with Energy Paper 59

Table 4.7	Transport Sector Growth Rates EP65 v EP59	
	1990 - 2000	1990 - 2020
EP65 Range	1.2% - 1.9%	1.3% - 2.2%
EP59 Range	1.0% - 2.8%	1.4% - 2.9%

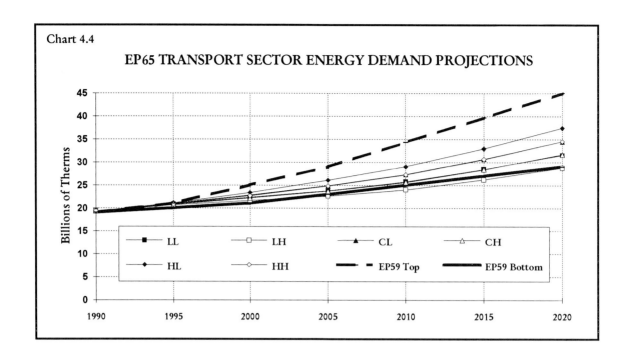

4.39 Chart 4.4 shows the transport sector projections and, for comparison, the top and bottom of the EP59 range. It can be seen that the new projections are located in the bottom half of the EP59 range and that the low GDP / high energy price scenario is marginally below the EP59 range. This difference in the projections reflects the inclusion of 1994 CCP measures. The impact of these measures is however partly offset by the lower oil price assumptions used here than in EP59.

4.40 The difference between the two sets of projections can also be explained by a move to more explicit modelling of diesel cars. The new projections now assume a larger switch away from petrol cars than was assumed for EP59; as diesel engines are more efficient than petrol engines this tends to lower fuel demand. A further factor reducing the EP65 road transport demand in the long run is the explicit modelling of household car ownership and its eventual saturation.

TOTAL FINAL USER ENERGY PROJECTIONS

4.41 Previous sections of Chapter 4 have detailed each sector's energy demand; this section aggregates them and compares them with EP59.

Table 4.8	Total Final Users				Billions of Therms		
	1990	1995	2000	2005	2010	2015	2020
LL	58.4	61.4	64.3	67.3	70.0	73.3	77.4
LH	58.4	60.7	62.5	64.9	66.9	70.4	74.3
CL	58.4	61.7	65.9	70.0	73.5	78.0	83.3
CH	58.4	61.1	64.1	67.6	70.6	75.2	80.4
HL	58.4	62.2	66.9	72.0	76.3	81.9	88.2
HH	58.4	61.6	65.1	69.6	73.4	79.0	85.4

4.42 Total final user energy demand is expected to grow by 0.7 - 1.4% per annum until 2000. For the period 1990 - 2020 (shown in Table 4.8) energy demand is projected to grow by 0.8 - 1.4% per annum. Table 4.9 shows which sectors are expected to grow fastest.

Table 4.9	Final User Sector Growth Rates Per Annum	
	1990 - 2000[3]	1990 - 2020
Domestic	- 0.2% ~ +0.1%	0.0% ~ +0.2%
Iron & Steel	- 0.3% ~ +1.2%	0.0% ~ +0.7%
Other Industry	- 0.5% ~ +0.8%	+0.4% ~ +1.2%
Service	+1.0% ~ +1.4%	+0.8% ~ +1.2%
Transport	+1.2% ~ +1.9%	+1.3% ~ +2.2%
Total Final User	+0.7% ~ +1.4%	+0.8% ~ +1.4%

4.43 It is apparent that the domestic, iron and steel and other industry sectors' energy demands all grow at a slower rate than total final user energy demand. Total final user energy demand growth during the projection period is therefore increasingly determined by energy demand growth in the service and transport sectors.

---

[3] 1991 - 2000 and 1991 - 2020 has been used for the domestic sector comparisons for the reasons explained in the domestic sector text.

Comparison with Energy Paper 59

4.44 Because of the changes in data definition that have occurred between the new and EP59 projections it is difficult to make precise comparisons between these two.[4] These differences aside however, Chart 4.5 shows the EP65 total final user energy demand projections and the EP59 range.

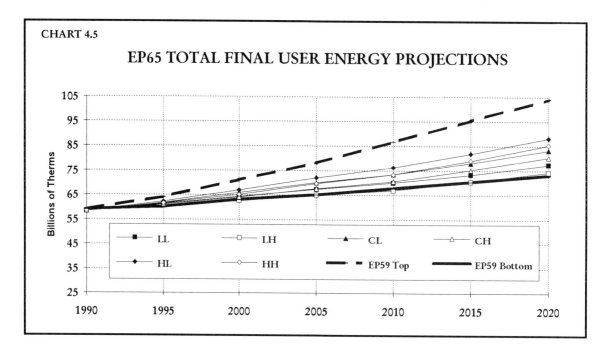

4.45 Two features stand out in Chart 4.5. First, the new projections are much closer to each other than the EP59 projections. This feature can be explained by the saturation and stock effects, detailed earlier on, acting to reduce the top of the range. These stock and saturation effects were weaker and less explicit in the EP59 models. The low fuel price scenarios further compress the range of energy projections by increasing the bottom of the range. As described in Chapter 2, a higher level of oil prices was assumed for EP59 than is assumed here.

4.46 The second feature is that the projections are generally lower than the EP59 projections. This is partly because the 1994 CCP measures reduce the projections relative to their EP59 counterparts. It is also partly because of the stock and saturation effects described above. However part of the reduction is undoubtedly due to the change in the way fossil fuels used for own-generation are recorded in the statistics. When allowance is made for this, total final energy demand in the period to 2000 lies in the top half of the EP59 range.

---

[4] See the other industry section of Chapter 4 for more details.

TOTAL FINAL USER ENERGY PROJECTIONS BY FUEL

4.47 This section examines the final user demand projections of electricity, gas, solid fuels and petroleum products.

Electricity Demand Projections

4.48 The projections of electricity demand shown in EP59 were for publicly supplied electricity and did not therefore include the demand for own-generated electricity from condensing steam turbines and CHP units outside the electricity supply industry[5]. The change in the treatment of fossil fuels for own-generation and the associated electricity means that the new projections of electricity demand include the demand for both public and own-generated electricity. Other things being equal this will increase recorded electricity demand. In order to allow for this the new electricity projections can be compared with EP59 in terms of percentage growth rates. For the same reason the fossil fuel projections are also compared in growth terms.

Table 4.10		Electricity Projections			Billions of Therms	
	1990	1995	2000	2005	2010	2015	2020
LL	9.4	10.0	11.1	11.9	12.4	12.8	13.4
LH	9.4	10.0	10.8	11.5	11.9	12.6	13.3
CL	9.4	10.1	11.4	12.4	13.0	13.8	14.7
CH	9.4	10.1	11.1	12.0	12.6	13.6	14.6
HL	9.4	10.1	11.5	12.7	13.5	14.5	15,6
HH	9.4	10.2	11.2	12.3	13.1	14.3	15.5

4.49 Table 4.10 indicates that electricity demand is projected to increase throughout the projection period. Relative to the 1990 level, electricity demand is expected to rise by 15 - 23% by 2000 and 42 - 66% by 2020. A comparison of the per annum growth rates for the period 1990 - 2000 (1.4 - 2.1%), with those for 1990 - 2020 (1.2 - 1.7%) shows that the growth rate is expected to decline somewhat in the long-run. This reduction in the growth

---

[5]Instead the fuel which generated this electricity was itself treated as an end-use. Following a change in statistical conventions, this fuel is now treated as an intermediate use, and the electricity from it is recorded as part of final consumption.

rate reflects saturation effects in the domestic and service sectors primarily, but also those in the other final user sectors.

Comparison with Energy Paper 59

Table 4.11	Per Annum Electricity Growth Rates EP65 v EP59	
	1990 - 2000	1990 - 2020
EP65 Range	1.4% - 2.1%	1.2% - 1.7%
EP59 Range	1.1% - 2.0%	1.9% - 3.5%

4.50 Table 4.11 indicates that while the new electricity demand growth rates for 1990 - 2000 are similar to those in EP59, they are significantly below them in the longer run. Most of this difference in growth rates arises from lower service sector electricity demand in 2020, particularly in the other appliances and air conditioning end-uses.

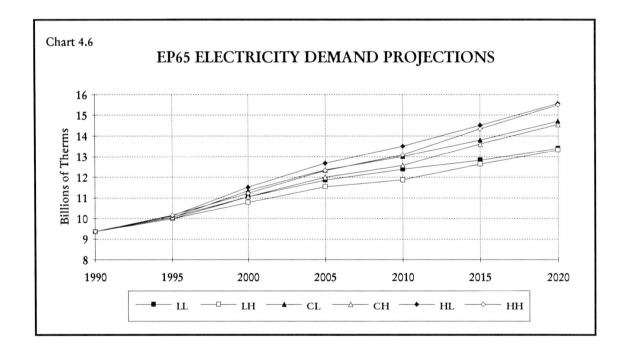

Gas Demand Projections

Table 4.12		Gas Projections			Billions of Therms		
	1990	1995	2000	2005	2010	2015	2020
LL	18.3	19.8	21.9	23.6	24.6	25.3	25.9
LH	18.3	19.9	22.2	24.0	25.1	25.8	26.2
CL	18.3	19.9	22.2	24.2	25.6	26.7	27.6
CH	18.3	20.0	22.6	24.6	26.0	27.0	27.7
HL	18.3	20.0	22.4	24.6	26.3	27.6	28.6
HH	18.3	20.1	22.8	25.0	26.6	27.9	28.7

4.51 The new projections suggest that final user gas demand will be some 20 - 25% higher in 2000 and 41 - 57% higher in 2020 than in 1990. Gas demand is projected to be marginally higher in the high energy price scenarios than in the low energy price scenarios. This is the opposite of what would be expected and it arises because of the large increase in oil prices between the two energy price scenarios, which improves the competitiveness of gas (and coal). Most of the switch out of oil is into solid fuels, hence the rather small increase in gas demand relative to the low energy price scenarios.

4.52 Approximately 4 billion therms of the projected increase in gas use of approximately 8 - 11 billion therms by 2020 shown in Table 4.12 comes from increased demand in the domestic sector. The remaining increase in gas demand comes from the service and industrial sectors, approximately 3 billion therms each.

Comparison with Energy Paper 59

Table 4.13	Per Annum Gas Growth Rates EP65 v EP59	
	1990 - 2000	1990 - 2020
EP65 Range	1.8% - 2.2%	1.2% - 1.5%
EP59 Range	0.0% - 1.0%	0.0% - 0.5%

4.53 Recent inter-fuel modelling by the DTI and the development of a new model of boiler and CHP demand[6] has led to the upward revision in projected gas demands. Consultations with some of the larger oil companies has also been instrumental in revising the long-term view of gas markets. In the short term to 2000, Table 4.13 and Chart 4.7 indicate that gas consumers are likely to continue to switch from other fuels into gas. Beyond the year 2000 however the possibilities for further switching to gas become more limited and market saturation, especially in the domestic sector, increasingly reduces the potential for further increases in gas demand.

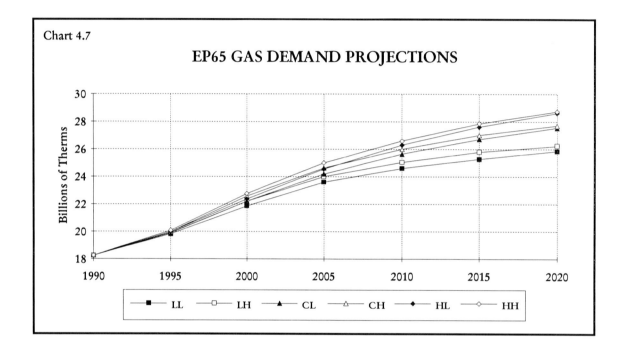

4.54 It should also be noted that a substantial part of the increase in gas demand between 1990 and 1995 is due to the low 1990 domestic sector gas demand, associated with a warm winter in that year, discussed previously.

---

[6] Science Policy Research Unit Heat and Power Model, volumes 1 - 5, Skea and Sorrell, University of Sussex, 1993.

Petroleum Product Demand Projections

Table 4.14		Oil Product Projections		Billions of Therms		
	1990	1995	2000	2005	2010	2015	2020
LL	25.1	26.8	27.1	28.0	29.6	32.2	35.3
LH	25.1	26.1	25.3	25.5	26.4	28.3	30.6
CL	25.1	26.9	27.8	29.4	31.3	34.4	38.2
CH	25.1	26.2	26.0	26.8	28.2	30.7	33.7
HL	25.1	27.1	28.5	30.6	33.0	36.8	41.3
HH	25.1	26.4	26.6	28.1	29.9	33.0	36.7

4.55 The EP65 oil product projections shown in Table 4.14 suggest that final user demand is projected to increase by 1 - 13% in 2000 when compared to 1990. In the longer term, final user oil demand is projected to increase by 21 - 64% when 2020 and 1990 are compared. Both income and prices can be seen to play important roles in determining the oil product projections. A comparison of the transport sector and oil product projections shows the sensitivity of transport growth as a determinant of oil product demand.

Comparison with Energy Paper 59

Table 4.15	Per Annum Oil Product Growth Rates EP65 v EP59	
	1990 - 2000	1990 - 2020
EP65 Range	0.1% - 1.3%	0.6% - 1.7%
EP59 Range	1.1% - 2.8%	1.0% - 2.5%

4.56 It is apparent from Table 4.15 that the projected per annum increase in oil product demand is substantially less in EP65 than in EP59. Two factors explain this lower growth. Firstly, as was discussed above, gas demand is somewhat higher in EP65 than in EP59. This implies that gas must be substituting for at least some of the oil shown in the EP59 projections. The second factor explaining this lower oil product demand growth is the increase in the real road transport fuel duties as part of the 1994 CCP measures.

4.57 The 5% per annum real increase in fuel duty helps to restrain oil demand growth up to the year 2000. However, the assumption that the duty is held constant in real terms beyond 2000 means that growth of oil product demand resumes after the year 2000. This can be clearly seen in Chart 4.8.

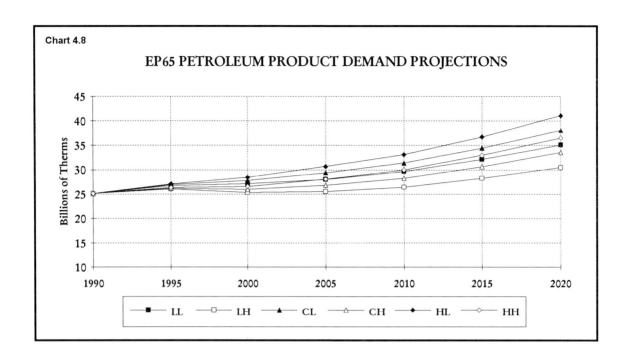

Solid Fuel Demand Projections

Table 4.16		Solid Fuel Projections			Billions of Therms		
	1990	1995	2000	2005	2010	2015	2020
LL	5.7	4.8	4.2	3.8	3.3	3.0	2.8
LH	5.7	4.7	4.2	3.8	3.5	3.7	4.1
CL	5.7	4.9	4.5	4.1	3.5	3.1	2.8
CH	5.7	4.8	4.4	4.1	3.8	3.9	4.4
HL	5.7	5.0	4.5	4.1	3.4	3.0	2.7
HH	5.7	4.9	4.5	4.2	3.8	3.8	4.4

4.58 Table 4.16 and Chart 4.9 show the current solid fuel demand projections. These suggest that solid fuel demand will continue its long-term decline. By the year 2000 solid fuel demand is projected to be some 20 - 26% lower than in 1990. Beyond the year 2000 this trend decline continues in the low prices scenarios; by 2020 it falls by about 52% of its 1990 level. After the year 2010 solid fuel demand recovers in the high energy price scenarios, ending about 25% below its 1990 level by 2020. The relative energy prices assumed and the extent to which consumers are influenced by coal's cost advantage in the high price scenario are therefore crucial to determining the extent of the likely future decline in final user solid fuel demand.

4.59 An examination of the detailed sectoral solid fuel demand projections shown in Annex A shows that during the period 1990 - 2020 the domestic sector's demand is projected to fall by approximately 1.5 billion therms, while the service sector's demand for solid fuel drops by around 300 million therms. It is only in the industrial sectors therefore that solid fuels are able to compete effectively with other fuels. In the iron and steel sector, for example, solid fuel demand either stabilises at its 1990 level or declines only slightly. In the other industry sector the change in solid fuel demand by 2020, compared to 1990, is -1 billion to +600 million therms. This increase in other industry's solid fuel demand in the high energy price scenarios goes some way to offsetting the decline in demand in the non-industrial sectors.

Comparison with Energy Paper 59

Table 4.17	Per Annum Solid Fuel Growth Rates EP65 v EP59	
	1990 - 2000	1990 - 2020
EP65 Range	-1.8% ~ -2.3%	-0.7% ~ -1.4%
EP59 Range	-1.6% ~ -2.9%	-1.3% ~ -1.5%

4.60 For the period 1990 - 2000 the current solid fuel projections suggest a smaller range of decline in demand per annum than that shown in EP59. Over the whole of the projection period, 1990 - 2020, the current projection range is much wider than the EP59 range. The wider range reflects different relative energy price assumptions and improvements to the energy models.

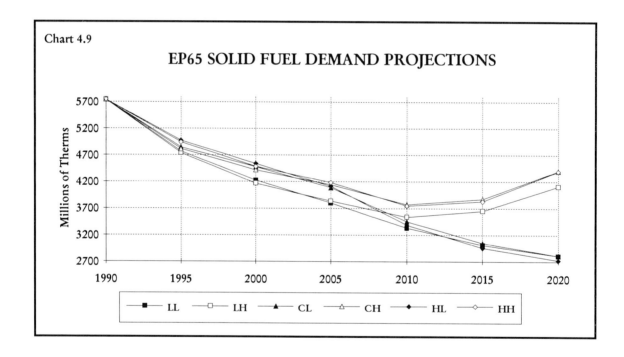

FINAL USER ENERGY DEMAND RATIOS

4.61 This section examines the energy ratios taken from the EP65 projections. On a sectoral level, the energy ratio shows the relationship between energy demand and output for any sector. At the national level it shows the ratio of total final user energy demand to GDP. Changes in the energy ratio can be loosely interpreted in terms of changes in the efficiency with which energy is, though changes in economic structure matter also. Changes in efficiency can arise from changes in fuel mix or technology. The ratios reported here are based on each sector's total energy demand is divided by that sector's output or income index. This ratio is then normalised so that in 1990 it is equal to one. Changes in the ratio can then be compared against the 1990 value of one.

4.62 All of the energy ratios shown in the charts below are relatively constant for the period 1990 - 1995 and decline thereafter. This behaviour of the energy ratios between 1990 and 1995 can be mainly attributed to the recession that occurred in the early part of this period. Post-1995 output grows above trend until the first few years of the next century resulting in large declines in the energy ratios. Thereafter output grows at a similar rate to that experienced historically and the energy ratios therefore continue to decline at broadly similar rates to those experienced historically.

Domestic Sector

4.63 Chart 4.10 shows the path followed by the domestic sector energy ratio over the projection period for each of the six main scenarios. It should be noted that the domestic sector energy ratio has been based on data from 1991, unlike the ratio for all the other sectors, which use 1990 as their base year. This is because of the problems arising from the fact that 1990 was an unusually warm winter, leading to low domestic sector energy consumption. The winter of 1991, on the other hand, was very typical in terms of temperature and provides a more reliable base upon which to calculate energy ratios.

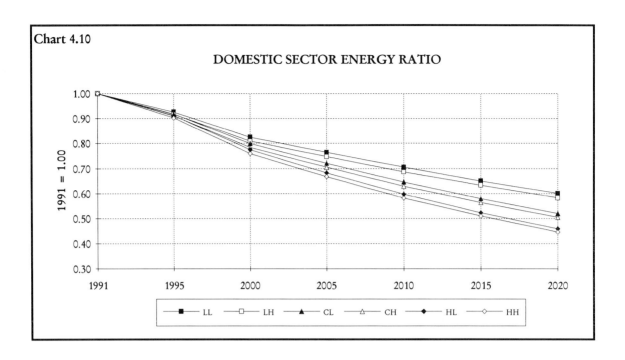

4.64 The chart shows that the ratio follows the same basic trend in all six scenarios. The ratio of energy demand to real personal disposable income falls in all scenarios between 1995 and 2020. By 2000 the ratio is projected to fall by 17 - 24% of its 1991 level and 40 - 55% by 2020. The main reason for the decline is a switch away from the less efficient fuels (solid fuels and oil) into gas. Thus as more households change from oil and solid fuel to gas as the means for heating their homes domestic energy use becomes more efficient. There is also likely to be some increase in efficiency due to technological advancement (such as improved insulation) which has a downward effect on the energy ratio.

Iron and Steel Industry

4.65 The energy ratio for the iron and steel industry is shown in Chart 4.11 below for all six scenarios.

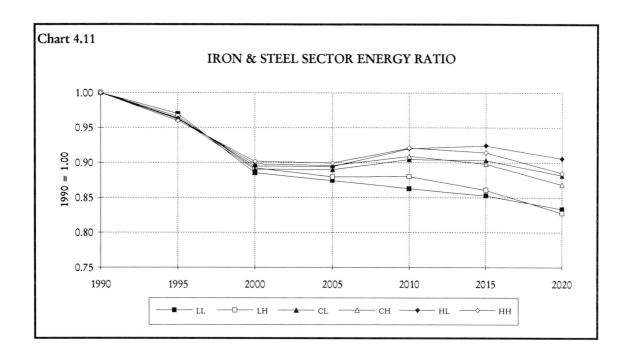

Chart 4.11 — IRON & STEEL SECTOR ENERGY RATIO

4.66 The iron and steel sector energy ratio, comparing energy demand with crude steel output, falls by 10 - 11% between 1990 and 2000 in all the scenarios. However, for the rest of the projection period, it only falls by any significant amount in the low growth scenarios (17% lower than the 1990 level in 2020). Indeed, in the central and high growth scenarios the ratio rises slightly (by 1 - 2% of its 1990 level) over some of the period between 2000 and 2020. There is very little change in the pattern of fuel consumption in this sector over the projection period and the ratio does not give evidence of any large-scale fuel switching.

4.67 The increase in the ratio beyond 2005 in the high and central GDP scenarios reflects higher EAF production. It also reflects higher output of finished products rather than just crude steel output.

Other Industry Sector

4.68 The other industry energy ratio, shown in Chart 4.12, follows distinct patterns in the two different fuel price scenarios. The ratio is calculated by dividing energy demand by an index of the sector's output. By 2000 it has fallen by 8 - 19% of its 1990 level. Between 1990 and 2020 it falls by 24 - 36%.

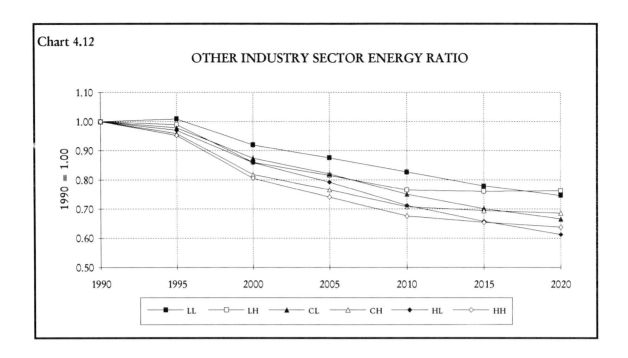

Chart 4.12 OTHER INDUSTRY SECTOR ENERGY RATIO

4.69 The difference between the two fuel price scenarios becomes apparent towards the end of the projection period. The ratio remains fairly constant between 2010 and 2020 in the high price scenarios, but continues to fall in the low price scenarios. Up to 2010 the ratio can be seen to be lower in the low price cases. The reason for this change is the relative price change of solid fuel compared with the other fossil fuels in the high price scenario. Solid fuels are relatively cheaper than the alternatives as all prices rise, especially towards the end of the projection period. This is because their price rises by much less than that of gas or oil. As was mentioned earlier, solid fuels are often less efficient than gas. So, although the high fuel prices lead to a decrease in the demand for **useful** energy, the switch from gas to coal offsets the impact on **delivered** fuel demand. This result must be regarded with some caution. If coal conversion efficiencies improve, relative to other fuels, in the future, a switch from gas to coal would be less likely to cause an increase in total delivered energy. In this case, the energy ratio would be more likely to continue its trend decline beyond 2010.

Service Sector

4.70 Chart 4.13 shows the energy ratio for the service sector, calculated using projected energy demand and an index of service sector output. This follows a very similar path to the domestic sector ratio, although the fall in the service sector ratio over the projection period is not so dramatic. There is very little change in the service sector energy ratio between 1990 and 2000 (a fall of 3 - 7%). However, by 2020 it has fallen by 25 - 39% from its 1990 level. The decline in the ratio can again be largely attributed to a switch from solid fuels and oil into gas. It can also be attributed to the saturation effects mentioned previously.

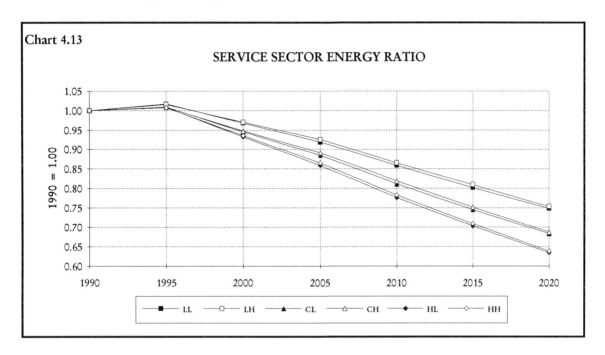

Chart 4.13 SERVICE SECTOR ENERGY RATIO

Transport Sector

4.71 The transport sector energy ratio is calculated using the total final expenditure series as an output index. Chart 4.14 shows that the ratio falls slightly over the projection period (by 4 - 9% between 1990 and 2000, and by 5 - 21% between 1990 and 2020). Most scenarios can be seen to follow the same basic downward path over the period.

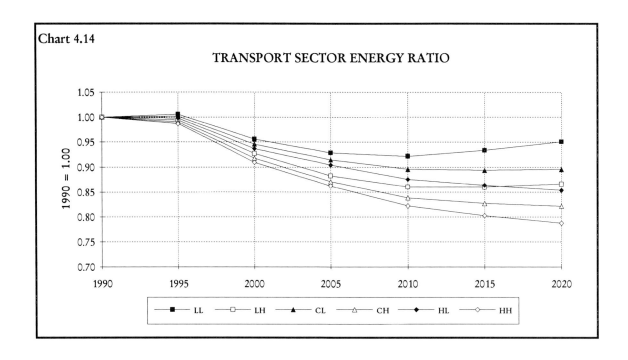

Chart 4.14 TRANSPORT SECTOR ENERGY RATIO

Total Final User Energy Demand/GDP Ratio

4.72 The ratio comparing total final user energy demand with GDP follows a broadly downward trend, as shown in Chart 4.15 below.

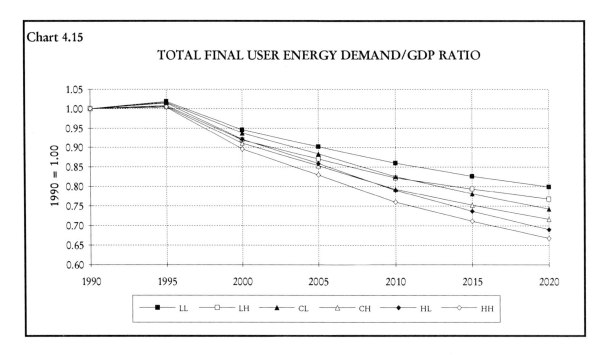

Chart 4.15 TOTAL FINAL USER ENERGY DEMAND/GDP RATIO

4.73 The ratio is expected to have risen between 1990 and 1995 - in part reflecting the warm winter of 1990 and in part the effects of the recession. Beyond 1995, the ratio reverts to its long-run trend. The long-run decline in the ratio is greatest in the high GDP growth cases. The ratio falls as fixed energy overheads are being employed more efficiently with the higher output in all sectors and as turnover in the capital stock is faster. The ratio is also lower in the high price scenarios as energy users try to use energy more efficiently.

CHAPTER 5. THE ELECTRICITY SUPPLY INDUSTRY[1]

BACKGROUND

5.1 Many of the trends identified in EP59 have continued in the three years since it was published. The predominant feature of the first half of the 1990s has been the surge in interest in CCGTs. From zero capacity in 1990, existing and committed CCGT capacity (plant already operating or under construction) in the ESI is now around 15 gigawatts (GW). In addition, there are many projects with consent to proceed. It is likely, therefore, that actual CCGT capacity in 2000 will exceed the current committed capacity. The amount of CCGT capacity is inextricably related to the amount of existing coal capacity. Most of the older, less efficient coal plants have now been closed, or closure plans have been announced. Some of the remaining coal plants have already been forced to run at significantly lower load factors than in recent years as the new CCGT plants have taken up a significant amount of baseload generation. The reduction in coal plant load factors has been accompanied by new coal contracts for much smaller tonnages of coal than previously.

5.2 At least some of the build of CCGTs has been due to a desire on the part of the regional electricity companies (RECs) to diversify their sources of electricity, or in some cases, to establish a foothold in the electricity generation market. These features make the short to medium term projection of CCGT and (predominantly) coal capacity difficult. In the longer term, the amount of CCGT (or other new plant build) is likely to be more closely related to changes in electricity demand.

---

[1] The ESI is defined here as including only those companies whose main business is to supply electricity. There are numerous other companies who also generate electricity, but whose main business is not electricity supply. Some additional CCGT plant (usually in CHP form) is being built by these companies. Such plant is modelled in the industrial sector.

ASSUMPTIONS

Plant Closures

5.3 Partly because of the reasons discussed above, it is difficult to be at all certain about the remaining lifetimes of existing coal and other plants. The approach taken in the projections has been to assume the closure by 2000 of all plants already announced as likely closure candidates by the generators themselves. In addition, one or two remaining smaller coal plants have been assumed to close. Thus by 2000, coal plant capacity, virtually all with 500MW, or bigger, sets, is assumed to amount to around 27GW. Although oil plant capacity is set rather lower in 2000 than in 1990, there must be doubts over whether even this amount of oil capacity is likely to survive to 2000 and beyond. This is particularly so, given the increase in duty in fuel oil in the 1994 budget. On the other hand, in the low price scenarios, oil remains very competitive against coal.

5.4 Notwithstanding the uncertainty about coal plant lifetimes, it is possible or even likely, that some plant lifetime extensions will take place. This could take the form of reconfiguring the operation of specific plant sets or perhaps the whole operation. For example, it is not technically difficult in many cases either to convert conventional coal fired plants to run on different **fuels** or to change plant operation to alternative **means** of electricity generation, such as fluidized bed or combined cycle operation. In the projections, the approach taken has been simply to allow, at minimal cost, some existing coal plants to continue operation for around 10 years longer than would otherwise be expected. In doing so, account has been taken of both the age and likely competitive position of individual plants.

5.5 The timing of nuclear plant closures is no less difficult. For some plants, a degree of life extension has been assumed. The effect of this is to increase the amount of nuclear capacity, relative to EP59, by around 1 - 2GW in the period up to 2010. The evidence to the Nuclear Review suggests a wide range of possible closure dates for plants in both England & Wales and Scotland. If anything, such evidence points to the possibility of capacity in Scotland remaining open for a little longer than has actually been assumed for the purpose of these projections.

5.6 The effect of the closure programme and other plant lifetime assumptions can be seen in Chart 5.1, which shows, for the CL scenario, the difference in plant capacities between the current projection and EP59. In the period up to 2000, the decrease in coal capacity and increase in CCGT capacity roughly match each other. Thereafter, coal capacity continues to fall, at a slower rate, partly reflecting the fall in load factors on some coal plants through time, with associated earlier closure prospects. Oil capacity is also a little lower than in EP59.

5.7 The small increase in nuclear capacity to 2010 is due to an assumption of slightly longer operating lives.

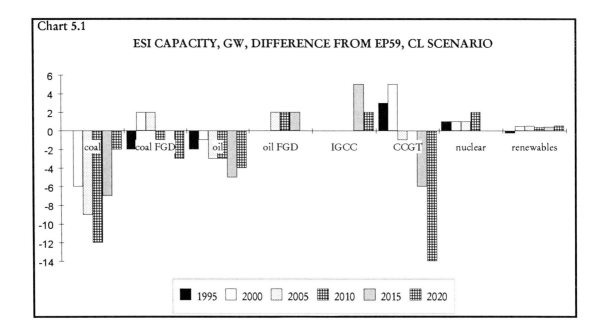

New Plant Capacity

5.8 The following section describes the various different types of new generating plant included in the modelling process as options for meeting electricity demand. It is assumed that the most cost-effective plant (i.e., that which generates at least cost - subject to various constraints, such as those on gaseous emissions) - will be built; this will obviously depend on what fuel prices are assumed. In the real world, when deciding what plant to build, generators have to take into account uncertainty about future fuel prices. In the face of uncertainty, it has been assumed for the purpose of the projections that a relatively diverse mix of plant will be maintained. This has been achieved in the model by limiting the share of any single input fuel to no more than two thirds of total ESI generating capacity.

5.9   CCGTs are currently the most popular choice for new capacity. Even since 1990, big advances have been made in CCGT technology, including both higher thermal efficiency and improvements in NOx control. Further improvements can be expected.

5.10   But other types of plant are beginning to gain recognition as possible contenders for future use in the UK. This section describes in some detail the range of current and prospective generating technologies, upon which the new projections are based.

5.11   One which has received considerable attention of late is the integrated gasification combined cycle (IGCC). This process, which is probably best described as being still in the demonstration stage, relies on gasifying the fuel input in order to produce a cleaner fuel for combustion. There are many variants of this process, but in each oxygen is added to the raw fuel (coal, oil products, refinery residuals, etc) and partially combusted to produce a raw hot gas. Some contaminants will be rejected at this stage-for example, for a raw coal fuel, most of the non-carbon material present melts and flows out of the (combustion) vessel, eventually forming a sand-like slag, thought to be non-leaching in character. Particulates, sulphur and other impurities are removed from the raw gas before it is passed to the gas turbine. The synthesis gas is then fed into a conventional combined cycle turbine system. The major difference between IGCC and CCGT technologies are the initial gasification and cleaning processes, both of which require energy, and thus reduce the efficiency of an IGCC compared with a CCGT. Because additional hardware is required, IGCC capital costs are also rather higher than for CCGTs. On the other hand, with appropriate design work, IGCC is capable of using many different types of fuel as input.

5.12   Other types of plant include coal and oil plant with FGD abatement equipment in-built. The capital costs of these plants are not dissimilar from those of IGCC, but the latter probably has a little more flexibility and is likely to turn out to be rather more thermally efficient. In terms of non-CO_2 abatement, the IGCC scores well, with near total removal of sulphur and substantial removal of NOx. If the more conventional types of plant were required to abate sulphur to the same degree, operating costs would rise substantially, thus reducing competitiveness.

5.13   A range of renewable technologies has been included in the modelling. Data for these technologies have been based on work by the Energy Technology Support Unit[2]. For some technologies, data has been updated to reflect more recent evidence. The Government's policy is to work towards 1500MW of new renewable capacity in the year 2000. The making of NFFO[3] orders is the Government's main instrument for achieving this. Thereafter, it is assumed that no further NFFO orders are made, so that the amount of any further build is solely determined by market forces.

5.14   One other type of technology, which has not attracted the same coverage as the CCGT, is the latest vintage of peaking turbine. This is sometimes referred to as the 'heavy frame' or 'heavy duty' turbine (so called to distinguish it from the earlier versions running on gasoil). The heavy duty gas turbine (HDGT) is essentially a CCGT without the back-end heat recovery plant and steam turbine which is used in a CCGT to boost thermal efficiency.

5.15   The modern HDGT is a much more efficient descendant of the version built either as a free-standing unit or attached to coal and oil-fired stations, mainly in the 1960s and 1970s. The notional efficiency of those plants was probably no more than 25% or so, but in practice, their short running times effectively reduced this to little more than 10%. The most recently announced HDGT is capable of generating electricity at an efficiency of around 34 - 35%, on a gross and after own-use basis. This technology is also capable of delivering electricity with low emissions - mainly due to the development of dry low-NOx emission systems. As with the CCGT, there are insignificant sulphur emissions from this type of plant. The HDGT has yet to be taken up in the UK, but the technology has already been ordered in other countries, most notably the United States.

5.16   Nuclear plant has been included, based on PWR technology. Prospects for this plant type have been tested at two capital cost levels-at £1300 per kilowatt and at £1000/kW (overnight costs - e.g., as if the plant were constructed overnight, thus incurring no interest charges - and at 1990 price levels).

---

[2] This work is described in New and Renewable Energy: Future Prospects in the UK, Energy paper 62, HMSO, March 1994.
[3] Non-fossil fuel obligation

Emissions Abatement

FGD Plant

5.17   It is assumed that FGD is retrofitted to Drax, Ratcliffe, Ferrybridge C and Longannet. In addition, FGD is assumed to be fitted to Pembroke, which is assumed to run on Orimulsion or a similar fuel from around the turn of the century.

Low-NOx Burners

5.18   In line with company plans, all major coal plants are assumed to have been fitted with low-NOx burners, by 1998.

The Generators' Improvement Programmes

5.19   Under HMIP plant authorizations, ESI plant operators are required to pursue a programme of improvements to the environmental performance of their existing plants. In the period during which these projections were being prepared, the details of these programmes had not been agreed and assumptions were necessary to reflect the likely outcome. What has become clearer recently is that retrofitting FGD to ageing coal or oil plant is not likely to be a favoured route for abating sulphur, despite the likely fall in cost of FGD plant. This has in part been influenced by the large build of CCGTs which has effectively reduced the scope for coal plants operating at baseload. At low load factors the economics of retrofitting FGD are poor compared with other alternatives.

5.20   Other techniques or operational policies to reduce emissions are available, the most obvious being to operate plant on lower load factors than they would otherwise be operated at. Although this reduces the output from the two major fossil-based generators' plants, there are no capital costs to bear and it may therefore be the most attractive short-term means of satisfying HMIP's requirements. This is the means by which the model has been constrained. An overall maximum load factor restriction has been placed on coal and oil plants, except those fitted with FGD units.

The load factor restriction was set at the levels indicated below:

Table 5.1:
Load factor restrictions on ESI plants

	1995	2000	2005-2020
coal plants	nr	55%	40%
oil plants	40%	40%	40%

nr = unrestricted

ESI Fuel Input Price Assumptions

5.21 As noted in the ESI methodology section, a supply curve for coal inputs to all major coal-fired power stations has been established. Oil products prices to the ESI correspond closely to the prices used by other consumers, with adjustments for the bulk purchasing power of the generators. Gas prices for existing customers and those with gas contracts already signed are set according to the escalation clauses which are contained in the contracts. New gas prices to the ESI move closely in line with the beach prices described in chapter 3. Orimulsion prices have been set at around the same level as landed coal prices.

ELECTRICITY SUPPLY INDUSTRY PROJECTIONS

New Plant Choice

5.22 Against a background of sizeable growth in electricity demand and particularly given that a large amount of existing capacity is retired from service between 2000 and 2010, there is a need for a considerable build of capacity in all scenarios. But, unlike the EP59 projections, where electricity demand increased substantially, total plant capacity does not increase dramatically. Table C1 in Annex C shows total ESI capacity in all of the six main scenarios and shows the corresponding figures for EP59 for purposes of comparison. Table C2 shows a breakdown of total ESI capacity in each of the six main scenarios.

5.23 In the low price cases, new plant choice is predominantly CCGTs in the period to 2010, but thereafter is mainly IGCC, running mostly on Orimulsion. In the high price cases, CCGTs are again preferred in the early period, but IGCC plant becomes competitive from 2010 and dominates new build in the later years, running on either coal or Orimulsion. The limit imposed on the share of any one fuel in the total ESI fuel mix means that in low price cases, fewer CCGTs (and more IGCC units) are built than would be the case in unconstrained scenarios in high price cases the effect is mainly to bring forward and increase the scale of the IGCC plant build. The limit does not bite until 2010 at earliest and usually, 2015.

5.24 Though not shown in table C2, it is notable that in most of these scenarios a sizeable build of HDGTs is an economic way of meeting the need for new capacity in 2005 and in some, beyond 2005. This mainly reflects the construction of a large tranche of baseload plant in the 1990s, which leaves the mix of plant on the system imbalanced. Also, the closure of a significant amount of capacity after 2000 - some of which was operating at low load factors - reduces the availability of plant which was previously helping to meet peak demands. Thereafter, more baseload plant is built, pushing other plants down the merit order, some of which are able to operate relatively flexibly at lower loads and therefore help meet peak demands.

5.25 Growth in generating capacity declines in the longer term. Partly this is caused by a slight decline in the plant margin through time. A major effect is that the growth in electricity demand moderates significantly post 2000, and own generation in industry and services continues to expand, reducing the scope for ESI capacity growth. It should be noted that IGCC plant is not taken up within the ESI until at least 2005 and in most scenarios not until 2010. It is possible that this type of technology may be taken up initially outside the ESI, perhaps most obviously in the refinery sector, where low value residual oils could be used as fuel, but also on any site where existing electricity generating plant is likely to become due for replacement in the next 5 - 10 years. There are many locations where this is possible. To be competitive against CCGTs, an IGCC plant would need access to cheaper fuel inputs (including perhaps low transport costs), since IGCC capital costs are higher. This is why such plants may first appear in the refinery sector, or where access to a refinery is easy.

5.26 There is no projected build of new nuclear capacity beyond Sizewell B in any of the scenarios.

5.27 The projections have been based on IGCC technology making inroads only in the ESI sector. This is mainly because any penetration achieved in other sectors is likely to be small in relation to likely penetration in the ESI. Further consideration is to be given to this issue in advance of further projections work to assess more accurately the scope for, and effect of, market penetration outside the ESI.

Electricity Prices

5.28 Over the last couple of years, electricity prices (before VAT) to most consumer groups have fallen in real terms. This trend is likely to continue in the short term as a result of a number of factors:

- falling fuel input prices, particularly for coal;

- the impact of tighter regulatory controls on prices;

- the declining impact of the fossil fuel levy;

- the level of capacity in relation to demand may remain high in the short run.

Some of these influences will continue to near the turn of the century. By this time, however, electricity prices will tend to become more heavily influenced by movements in fuel input prices and perhaps by capacity charges, as total capacity and peak demand come more into balance.

5.29 Overall, electricity prices are projected to fall in real terms to the end of the century in all scenarios. Prices in the low energy price scenarios tend to increase a little thereafter, while prices in the high price scenarios tend to remain roughly constant at the 2000 level, though somewhat above those in the low price scenarios. One reason for the differential movements in electricity prices in the low and high price scenarios is that rather more sulphur abatement is required in the low price cases, increasing electricity generation costs.

The Demand for Fuels

5.30   In all scenarios, the demand for coal in the ESI remains substantially lower than historic levels. The large amount of CCGT capacity already committed, plus further gas build, reduces the scope for coal plant to run at high loads. This effect is reinforced by the restrictions assumed to be applied to the load factors on many coal plants as a result of improvement programmes to be agreed between the generators and HMIP. Nonetheless, coal retains a significant share of the ESI fuel market, particularly in the high price scenarios (oil and gas prices rise considerably in these scenarios, so that the price relativities favour coal). Table C3 shows ESI fuel burn in each scenario.

Coal

5.31   In the short term, coal demand is likely to be virtually all UK-sourced, reflecting the existing coal contracts. Over time, there is expected to be a greater reliance on imported coal as the coal contracts expire and as some UK reserves are exhausted, but the outcome will be heavily influenced by the extent to which the newly privatised coal mines are able improve their productivity. In high price scenarios, the demand for coal increases beyond 2005 as some of the newly constructed IGCC plant burns coal (mostly imported).

Gas

5.32   The demand for gas rises to around 2005 or 2010 in most scenarios, falling or remaining steady thereafter, either because it is displaced by coal and orimulsion (high price scenarios) or is constrained by the generators' desire to maintain a diverse mix of fuels.

Oil

5.33   Oil demand is low in the short term as the current coal contracts limit the scope for oil burn. In the low price scenarios, oil burn recovers by 2000 somewhat, thereafter falling as the impact of sulphur abatement measures begin to bite on (sulphurous) heavy fuel oil and as oil plants close. In the high price scenarios, fuel oil burn remains very low post-1995, due to an uncompetitive price.

Orimulsion

5.34   Orimulsion is an emulsified heavy oil, produced in Venezuela. There are many other sources of heavy oils and the term Orimulsion is used here to denote all such oils. Orimulsion burn increases post-2000 as Pembroke is assumed to be modified to accept the fuel. A further 500MW of existing, unspecified oil plant is also assumed to burn orimulsion at this time. A build of IGCC plant takes place after 2005, some of which runs on Orimulsion.

Nuclear

5.35   Nuclear fuel inputs move in line with capacity, declining post-2000 and forming only a very small proportion of total fuel inputs in 2020 (around 3-4%).

Renewables

5.36   Renewables capacity increases in the longer term, more so in the high price scenario. Nearly 3GW of capacity may be in place in 2000, rising to between 3.5 and 5GW in 2020.

COMPARISON WITH EP59

5.37   The chief feature of the ESI projections is that ESI capacity is considerably lower than in EP59. This is mainly due to lower electricity demand in total and lower market share for the ESI as own-generation - largely based on new gas fired capacity - is higher than in EP59. Total capacity is down by around 1 - 3GW in the low and central growth scenarios and by around 5GW in the high growth scenarios in 2000. By 2020, the reductions across the various scenarios range between 20 and 65GW. As a consequence, CCGT capacity and, in some scenarios, IGCC capacity is lower than in EP59 in the longer term.

CHAPTER 6. PRIMARY ENERGY DEMAND

BACKGROUND

6.1　Since the EP59 modelling exercise, there have been many changes to the definition and calculation of energy demand estimates. The main change affecting the modelling work reported in this chapter has been a revised method for calculating the contribution of nuclear, hydro and imported electricity to primary demand.

6.2　Details of these changes can be found in the consultation document issued in November 1993[1]. The methodological changes have changed the numerical estimates of the main measures of energy demand. For example, in 1992, primary energy demand under the previous methodology was 206.6 million tonnes of oil equivalent. Under the new methodology, it was estimated to be 220.0 Mtoe.

PROJECTIONS OF PRIMARY ENERGY DEMAND

6.3　Table D1 in Annex D shows the current estimates of primary energy demand for each scenario. Chart D1 shows the change in the projection of primary energy demand for each of the six scenarios compared to EP59. In the short term, EP59 underestimated the growth of primary energy demand from its 1990 level, particularly in the high price scenarios. This is in part because energy prices have generally turned out towards the bottom of the range assumed in EP59 and in part because the impact of 1990 being a warmer than average year was underestimated. Looking further ahead, there are relatively small changes in the primary demand estimates for the low and central economic growth cases, but rather more of a reduction for the high growth case. Chart 6.1 shows the current estimates for each of the six main scenarios.

---

[1] Review of Statistical Methodologies for the Compilation of Overall Energy Data, DTI, 1994. Copies may be obtained from the DTI by contacting Mike Ward on 0171 238 3576.

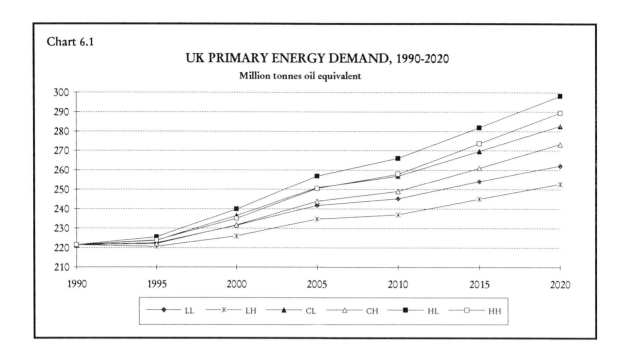

Chart 6.1

UK PRIMARY ENERGY DEMAND, 1990-2020

Million tonnes oil equivalent

6.4 Several points of interest emerge from Chart 6.1. First, there is considerable overlapping of energy demands across scenarios in the early years. For example, demand in the HH scenario is below that in CL up to and including 2005. This illustrates the point that, at least in the shorter term, the effect of the high price assumptions is sufficient to offset the effect of higher economic growth.

6.5 One other major feature of the projections is that the difference between the lowest and highest projected demands in 2020 is just less than 50 million tonnes. On the same definitional basis, the spread of demands in EP59 was around 125 million tonnes. The lowest demand in the current projections is around 250 Mtoe, compared with 256 Mtoe in EP59 - not very different. But the highest projection currently is just less than 300 Mtoe, compared with 380 Mtoe in EP59.

6.6 The reduction in the top of the range of energy demands is due to a number of factors. One is the overall reduction in electricity demand. This has the effect of reducing substantially the amount of fuel inputs required to generate electricity. The factors behind this have already been described in the sections on final energy demand and include the effects of saturation in some energy markets.

Primary Demand by Fuel

6.7 Table D2 shows the fuel breakdown of primary energy demand. The dominant features of this table are that the demand for gas and oil increases markedly in the long run (in the low price cases by an average-across all three scenarios-of around 130% and 65% respectively, between 1990 and 2020 and by 100% and 30% respectively in the high price cases). The increase in oil demand is shared between transport fuels and the use of orimulsion in the ESI. Gas and oil dominate total primary demand in the low price cases, while their dominance is less marked in the high price cases, reflecting the recovery in coal use after 2010.

PRIMARY ENERGY RATIO

6.8 Chart 6.2 shows the primary energy ratio for the LH and HL scenarios. These scenarios represent the extremities of the range of primary energy demands. The ratios for the low high and high low cases fall by around 1.2% and 1.4% per annum between 1990 and 2005 respectively. Between 2005 and 2020, the ratios fall by 1.3% and 1.8% per annum. These falls compare with 1% and 1.7% per annum in the first period and 0.8% per annum (for both cases) in the second period in the equivalent (low and high primary demand estimates) EP59 cases. The current projected fall in the energy ratio is thus within quite a narrow range for the period 1990 - 2005, and within the range for EP59, while for the period 2005 - 2020, the current projections show a more rapid fall in the energy ratio than in EP59.

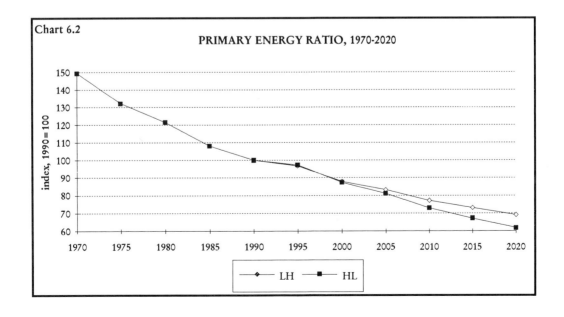

6.9 The energy ratio falls more rapidly in HL than in LH. This is largely because HL is more gas-intensive than LH in the ESI and other sectors. Due to its greater efficiency in use, compared with coal, this reduces energy use per unit of output. The high growth scenarios also reflect the increasing influence of saturation effects in some energy markets and of faster turnover in the capital stock, which improves the average efficiency of energy-using equipment.

CHAPTER 7. CARBON DIOXIDE EMISSIONS

BACKGROUND

7.1   Carbon dioxide emissions have been falling in recent years. The estimate for 1992 is 155 MtC. This is some 3 MtC below the 1990 level of 158 MtC, which forms the basis of the UK's stated target of returning emissions to their 1990 levels by 2000. The relatively warm winters of 1992 - 1994 have contributed to lowering the level of emissions in these years. The effect of the economic recession has also reduced energy demand, which in 1994 was only marginally above its 1990 level. The gradual increase in energy demand in 1993 and 1994 has been mainly based on fuels with zero, low or medium carbon contents, i.e., gas, oil and nuclear, with coal demand down by around 25% since 1990, mainly due to lower use of coal in the ESI.

PROJECTED CO_2 EMISSIONS

7.2   The emissions projections are compiled on an almost identical coverage to those in EP59. The single difference is that emissions from combustion of biofuels (in industry and power stations) were included in EP59, but are not included in the new projections. The difference which results from this increases through time, from less than ½ million tonnes of carbon in 1995 to around 2 million tonnes in 2020. The current set of projections are based on the conventions used by AEA Technology in calculating the official UK emissions inventory.

7.3   The carbon emissions estimates presented in Chart 7.1 include the effects on energy use of the Government's CCP and the subsequent modifications to it (e.g., the changes in fuel duties made in November and December 1994, in part to compensate for VAT on domestic fuels remaining at 8%). They also allow for the effects on carbon of assumed measures aimed at reducing sulphur emissions in the whole economy to levels agreed in the UN Economic Commission for Europe's (UNECE) Second Sulphur Protocol.

7.4 Table E1 in Annex E shows the aggregate figures[1]. As mentioned in the assumptions chapter, two more extreme energy price scenarios have been tested for the period up to 2000. The results for these two scenarios are dealt with at the very end of the CO_2 section of this chapter.

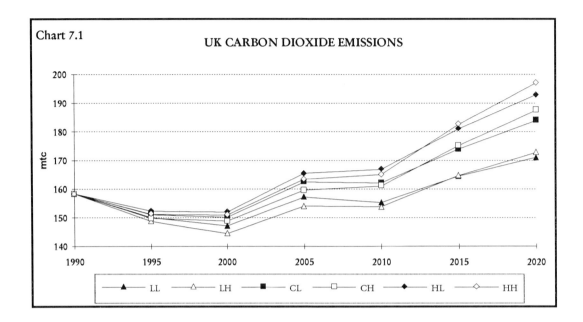

Chart 7.1 UK CARBON DIOXIDE EMISSIONS

The Effect of the Climate Change Programme

7.5 When it was published in January 1994 the Climate Change Programme estimated that the measures outlined in it would together save 10 MtC by 2000. Since the Programme was prepared, there have been some changes: VAT on domestic energy has been held at 8% and, partly to compensate for this, road fuel duties have been further increased. There have also been changes in energy markets and in energy industries, which will affect how much CO_2 the measures can be expected to save. For example:

- CCGT build in the ESI has been greater than expected in EP59. Measures aimed at saving electricity are therefore more likely to result in lower gas burn than in lower coal burn. Because gas has a lower carbon content than coal and can be burned with higher efficiency, this effectively reduces the carbon savings associated with any given

---

[1] The emissions projections presented in this chapter and in Annex E are based on aggregations of the detailed groupings to be found in the Department of the Environment's annual 'Digest of Environmental Protection and Water Statistics'.

fall in electricity demand. The results below are calculated on the basis that mainly gas-fired generation is displaced by the CCP; if instead coal-fired generation were displaced, the "with CCP" CO_2 projections, shown in Table 7.1, would be further reduced.

- expectations of future pre-tax fossil fuel prices have fallen. The transport fuel duty increases in the CCP therefore represent a larger percentage increase in final price than previously. This tends to increase the estimated CO_2 savings in the transport sector.

Table 7.1 UK CO_2 Emissions in 2000					MtC
	EP59 Projections[1] CL Scenario	EP65 Projections			
		CL Scenario		CH Scenario	
		without CCP	with CCP	without CCP	with CCP
1990(base)	158.3	158.3	158.3	158.3	158.3
2000	169.4	157.8	149.7	157.2	148.4

1: Adjusted to same basis as EP65.

7.6   Even in the absence of the measures in the CCP, Table 7.1 shows that other developments (notably the switch to gas in the ESI) have reduced the projected levels of CO_2 emissions - by about 12 MtC in 2000 in the two central growth scenarios. On the assumption that the non-fiscal measures are all successful in saving the amounts of energy set out in the CCP, the Programme as a whole would be expected to save a further 8 - 9 MtC by 2000.

7.7   There is, however, particular uncertainty about one measure in the programme. To date, the EST has been unable to obtain funds for the programmes originally intended, which would achieve in full the energy savings anticipated in the CCP. The associated carbon savings attributable to the EST in 2000 were originally estimated to be 2.5 MtC. The current projections suggest that, even if the EST achieved its energy saving target in full, this would reduce carbon emissions by only about 1.6 MtC. The Trust is currently reassessing its programme. It estimates that schemes for which it has already obtained funding can be expected to save 0.3 MtC. Were this situation to continue, the savings expected from the CCP as a whole would fall to 7 - 8 MtC in 2000, and there would be a commensurate rise in the "with CCP" emission estimates in Table 7.1.

General Trends in Emissions

7.8   In each scenario emissions are lower in the year 2000 than in 1990. Thus the target of returning emissions to their 1990 level in the year 2000 would be achieved in all the projections. In the longer term, emissions tend to rise again, as fuel switching (into gas) becomes less economic or more difficult (partly because of the imposed ESI fuel constraint), and as energy demand continues to grow. The drift upwards in carbon emissions is also partly due to the steady decline in nuclear output and the continued dominance of fossil fuels in the energy market. Carbon emissions stay below the 1990 level in the cental and high growth scenarios until between 2000 and 2005 and, in the low growth scenarios, until between 2005 and 2010.

Sectoral Trends

7.9   The reduction in emissions between 1990 and 2000 is dominated by reductions in the ESI. For example, in the CL and CH scenarios, the fall in ESI emissions between 1990 and 2000 is 18 and 16 million tonnes respectively. In the period to 2000, the fall in ESI emissions is caused by the switch into gas rising nuclear output and by the influence of the CCP. Table E2 presents sectoral emissions for each scenario. In this table, CO_2 emissions arising from the ESI are not allocated to end-users. The effect of redistributing the emissions arising in the ESI and refineries to the ultimate users of electricity and oil products is shown in Table E6. Each of the sectors is now discussed.

7.10   It should be noted that emissions from fuels used for own-generation of electricity are included in the relevant sector's emissions totals (i.e. industry or services).

ESI

ESI
7.11   In both the low and high energy price scenarios, ESI emissions fall sharply to 2000 but grow in the longer term. By 2020 in the low energy price scenarios, ESI emissions are close to, or remain below, 1990 levels. This latter feature does not hold so strongly in the high price cases, where emissions are at, or above, 1990 levels. The difference is that in the low price scenarios, there is a large amount of gas fired capacity , whereas high price scenarios have less gas fired capacity and IGCC plant runs on coal or Orimulsion, with their higher carbon contents.

Transport
7.12   Transport emissions rise steadily in all scenarios. In the low price scenarios, transport sector emissions are around the same size as those from the ESI. These two sectors combined account for the majority of emissions growth in the very long term.

Domestic
7.13   After increasing between 1990 and 1995 (mainly due to the effect of a very mild 1990 Winter on energy demand in that year), emissions in the domestic sector fall gradually, reflecting greater use of gas and less oil and coal. There is very little growth in overall energy demand in this sector.

Services
7.14   Emissions in the service sector increase modestly in all scenarios. Much of the growth in energy demand in this sector is of electricity. Emissions change little between the low and high price cases.

Refineries
7.15   Emissions from refineries increase modestly over time, reflecting to some degree a need for greater throughput, but also of extra processing requirements.

Other Industry and Agriculture
7.16   Other industry emissions increase a little to 2000 in most scenarios and in the longer term typically remain flat or fall slightly. In the CH and HH cases, there is enough energy demand and use of coal to produce a

modest increase in emissions in the longer term. There is very little difference in coal use for final demands in 2005, but thereafter, coal use increases markedly in the high price cases, but declines in the low price cases. Oil, and to some extent gas, lose out.

Emissions by End User

7.17   Emissions which arise in the course of producing secondary fuels, such as electricity, can be hypothecated to end-users to show the total emissions associated with their energy consumption. This makes a difference to the emission trends in some sectors (see Table E6). For example, instead of emissions remaining either flat or falling in the domestic sector (on a source basis), emissions rise somewhat post-2000. This is mainly due to the effect of allocating a part of ESI emissions (which increase post-2000) to the domestic sector, whose electricity demand increases over time. In the period up to 2000, domestic sector emissions on an end-user basis fall, reflecting the fall in ESI emissions over the same period. A similar phenomenon occurs in the commercial/public service sector (the category name employed in the national inventory - EP65 uses the collective term 'services', e.g., see table E2) and in industry.

The Increase in Emissions Between 2000 and 2005

7.18   A feature of these projections is the increase in emissions between 2000 and 2005 in every scenario. One reason for this is that in 2000, there is a large stock of CCGT plant which operates at baseload. Some of the older coal and other plants, running at low load factors, close post-2000, leading to an ex-ante deficiency of plant to meet peak demand. There is therefore a requirement for new generating capacity (since without it, there is an increased risk of there being insufficient capacity to meet demand). The cheapest means of meeting this requirement for capacity in 2005 appears to be to build peaking plant and to meet increasing baseload demand by running existing plants at higher load factors.

7.19   The consequence of this is that remaining coal plants operate at higher load factors than in 2000. Hence coal burn increases in every scenario, leading directly to higher emissions. The likelihood of such an outcome in the real world is unclear - much will depend on the success or otherwise of schemes to reduce peak demands; as the prospects for this are so unclear, little allowance has been made for reductions in peak demands. If peak demands can be reduced, then arguably coal burn would not recover to the degree portrayed here.

7.20   The fall in nuclear capacity between 2000 and 2005 is another cause of the increase in emissions between these two years.   The retirement of existing nuclear capacity and its replacement by gas, coal or Orimulsion is also a significant force acting to increase emissions in the long term, described below.

Influences on Longer-term Emissions

7.21   Up to around 2010, emissions are higher in low energy price cases than in high energy price cases because energy demand is higher. In the high price cases, coal and orimulsion prices are such that relative fuel prices favour an earlier build of IGCC plant in the ESI, running on fuels with high carbon contents. ESI fuel switching therefore becomes important in the high price scenarios, causing emissions in high price cases to exceed those in the low price cases post-2010.

7.22   To help illustrate the effect of the alternative price assumptions, carbon emissions by fuel source are presented in table E5. Compared with the position in 1990, emissions from coal use are greatly reduced by 2000 in all scenarios. Between 2000 and 2020, emissions from coal combustion decrease further in the low price cases but rise in the high price cases. Emissions from the use of gas increase rapidly up to 2000, followed by decelerating growth post-2000 as the penetration of gas into energy markets slows (partly due to the constraint applied in the ESI market). Emissions from combustion of motor spirit are basically flat throughout, while emissions from DERV increase significantly, reflecting its greater penetration in the transport fuel market. Emissions from the use of fuel oil (including orimulsion or other emulsified fuels) typically increase post-1995 (where use is low due to the limited short term scope in the ESI market, owing to the coal contracts). In the long-term, particularly, emissions from fuel oil increase somewhat, as IGCC plant is built in the ESI and burns Orimulsion.

7.23   Chart 7.2 shows how the different price scenarios affect sectoral emissions. In the high price case (CH) ESI emissions are higher because coal is more competitive than in the low price case (CL). The same is true in other industry, although this effect does not become clear until the end of the period considered; in earlier years the effect of high prices on overall energy demand in industry outweighs the effect of increasing coal use.

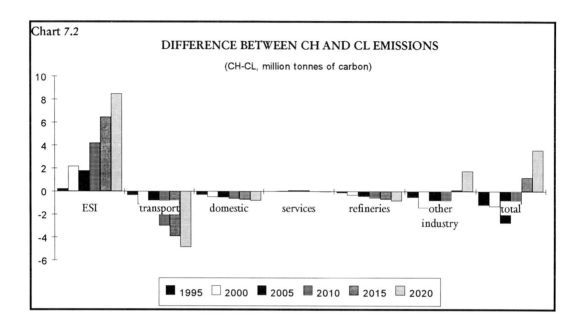

Differences Compared with EP59

7.24 Primary energy demand increases gradually in all scenarios. As has been pointed out in earlier sections of this report, the growth in demand for energy is now expected to be rather less than in EP59, particularly in the high economic growth cases. This reduction in energy demand cuts projected emissions directly. In the high economic growth cases, it is the reduction in energy demand projections compared with EP59 which is the main cause of the reduction in projected carbon emissions. In the central and low growth cases, the reasons for the reduction in estimated emissions varies according to price scenario. In low price scenarios, there is a less carbon intensive fuel mix, due to gas displacing oil and coal. In the high price scenarios, the fuel mix is not dissimilar from those in EP59, but there is rather less energy demand in total.

7.25 Lower ESI emissions account for the bulk of the overall change. ESI emissions are lower due to increased gas penetration in the short term (to 2000) and in the longer term, to lower electricity demand in total and also the impact of a growing amount of own-generation within industry and, to a degree, services. Charts E1 and E2 in Annex E show the change in CO2 emissions from EP59 for the CL and CH scenarios only. Other sectors contribute relatively little to the overall changes in emissions. Projected emissions from the domestic sector have increased as a result of improved modelling of temperature effects (see Chapter 4), though the trend in the current projections is still relatively flat. Industrial emissions are a little lower in the low price scenarios, reflecting higher penetration of gas, but slightly higher in the high price scenarios, reflecting a sizeable increase in final energy demand in this sector, in conjunction with a sizeable increase in coal use post-2005. It should be noted that emissions would be higher if it were not for the moderating effect of measures in the CCP. Transport emissions are somewhat below the levels reported in EP59, mainly due to the policy of real increases in fuel duties each year up to 2000 as part of the CCP.

Very Low and Very High Prices

7.26 To examine the effect on CO_2 of some extreme short to medium term price assumptions, the high economic growth assumption was combined with a very low price assumption and a low economic growth assumption was combined with a very high price assumption. These two scenarios are then more extreme versions of the HL and LH scenarios. The effect of these two extreme price scenarios is to increase CO_2 emissions in the HL case in 2000 by 3 MtC and to reduce LH emissions by almost 8 MtC. Emissions beyond 2000 and associated energy demands are not reported here, since such extreme prices are not thought likely to persist for more than a few years, though clearly price effects could still be working through the energy markets beyond 2000.

7.27 Annex G sets out energy demands and CO_2 emissions in these scenarios in more detail. Table G3 shows the existing top and bottom of the range of CO_2 emissions and primary demand in the year 2000 (HL and LH scenarios) with the two price variants for comparison.

PROJECTED SO2 EMISSIONS

Background

7.28   In order to ensure that the projections of carbon dioxide emissions are soundly based, it has been necessary to make assumptions about how HMIP policy on ESI plants might develop over time (set out in the assumptions section of Chapter 5) and about how the UK will meet its existing international commitment to reduce sulphur emissions to 30% and 20% of the 1980 level in 2005 and 2010 respectively. These assumptions are needed because these policies will also affect carbon emissions. However the assumptions do not pretend to be any more than indicative at this stage, ahead of work which will be undertaken through the course of this year and next in preparing a national sulphur strategy. DoE's National Atmospheric Emissions Inventory provides authoritative projections in fulfilment of the UK's European and international reporting obligations and will reflect work on the national sulphur strategy.

7.29   The projections have to take account of the UK's international commitments to abatement of sulphur emissions. For those scenarios in which emissions would otherwise exceed the target levels in 2005 (HH, LL, CL and HL) and for all scenarios in 2010, further constraints are therefore incorporated into the modelling. For simplicity, these are modelled as broadly market-based measures. However, the projected carbon emissions would not be significantly altered if alternative assumptions were adopted.

Projected SO$_2$ Emissions

7.30   The projected path of sulphur emissions in meeting the targets is shown in Chart 7.3. The sectoral breakdown for the CL scenario is shown in Table E3.

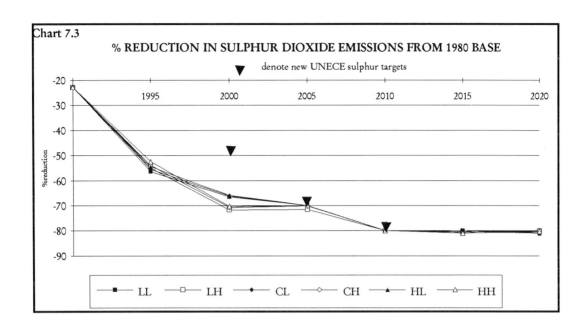

The chart shows a sharp fall in sulphur emissions occurring between 1990 and 1995. The reduction, averaging approximately 1½ million tonnes across all scenarios, mainly occurs in the ESI, due to reduced oil and coal burn, reinforced by the commissioning of FGD units at Drax and Ratcliffe. These factors continue to influence emissions to 2000 as further CCGTs are commissioned and the remaining FGD units become operational.

7.31   FGD units are assumed to be bid in at full marginal cost, which, since there are additional costs incurred in operating FGD units, decreases the competitiveness of these plants. As a result, in the low energy price scenarios, the FGD units collectively run at relatively low load factors, though the actual load varies somewhat between plants. In the high energy price scenarios, more favourable to coal, all FGD units run at high load factors.

7.32   Chart 7.3 reflects the result that there is more heavy fuel oil burn in the low energy price scenarios and hence more sulphur emissions in the year 2000. The amount of additional abatement required to achieve the 2005 and 2010 targets is relatively small and in two scenarios the 2005 target is achieved without any sulphur specific measures.

7.33   However, beyond 2010 the cheapest abatement measures have already been taken and the marginal costs of holding emissions at 20% of their 1980 level rise sharply. By this time, gas burn in the ESI is approaching its maximum tenable share and retrofitting FGD to any remaining conventional coal or oil plants is extremely unlikely due to the age of plant and low load factors on such plants. A further issue throughout is that the sulphur contents of coal and fuel oil commonly used in some sectors are similar, so that fuel switching to abate sulphur is not necessarily an easy option.

7.34   Around 2010 the main sector in which remaining sulphur levels are significant is industry, where further abatement is inhibited by a continuing presence of oil and coal in industrial energy markets, especially for oil in the low price scenarios (due to low fuel oil prices) and also to a lesser extent for coal, where a recovery takes place in coal use post-2005 in the high price scenarios.

7.35   The imposition of sulphur abatement measures to meet the targets for 2005 and 2010 also reduces carbon emissions. The reductions in carbon emissions in 2005 and 2010 vary according to scenario, but are in the range 1 to 4 million tonnes of carbon. They occur in most sectors, especially the ESI and industry, through fuel switching and energy demand reductions.

7.36   The United Kingdom also has a commitment to meet emission limits under the Large Combustion Plant Directive (LCPD). These limits, which vary by sector, are comfortably achieved in each scenario, with a significant amount of headroom in the ESI and other industry. Table E4 shows, for the CL scenario only, how emissions compare with the sectoral limits.   It should be noted that the target dates for meeting the LCPD are 1998 and 2003, rather than the years reported in table E4. But emissions are clearly below the limits, so further detailed examination of the annual path of emissions is considered unnecessary.

ANNEX A

FINAL USER ENERGY DEMAND BY FUEL AND SECTOR

DOMESTIC SECTOR
- Table A.1 - Final Energy Demand by Fuel - Low Fuel Prices
- Table A.2 - Final Energy Demand by Fuel - High Fuel Prices

IRON AND STEEL INDUSTRY SECTOR
- Table A.3 - Final Energy Demand by Fuel - Low Fuel Prices
- Table A.4 - Final Energy Demand by Fuel - High Fuel Prices

OTHER INDUSTRY SECTOR
- Table A.5 - Final Energy Demand by Fuel - Low Fuel Prices
- Table A.6 - Final Energy Demand by Fuel - High Fuel Prices

SERVICE SECTOR
- Table A.7 - Final Energy Demand by Fuel - Low Fuel Prices
- Table A.8 - Final Energy Demand by Fuel - High Fuel Prices

TRANSPORT SECTOR
- Table A.9 - Final Energy Demand by Fuel - Low Fuel Prices
- Table A.10 - Final Energy Demand by Fuel - High Fuel Prices

TOTAL FINAL USERS
- Table A.11 - Final Energy Demand by Fuel - Low Fuel Prices
- Table A.12 - Final Energy Demand by Fuel - High Fuel Prices

Some of the historic data shown in this annex revises previous estimates shown in the 1994 Digest of United Kingdom Energy Statistics.

TABLE A.1 - LOW FUEL PRICES: DOMESTIC SECTOR - FINAL ENERGY DEMAND BY FUEL

Billion Therms	1985	1990	1995	2000	2005	2010	2015	2020
Low GDP, Low Prices (LL)								
Solid Fuel	3.0	1.7	1.6	1.2	0.9	0.7	0.6	0.5
Oil	1.0	1.0	1.1	0.6	0.3	0.2	0.2	0.2
Gas	9.7	10.2	11.9	12.7	13.4	13.8	14.0	14.1
Electricity	3.0	3.2	3.4	3.5	3.6	3.7	3.8	3.9
Total	**16.7**	**16.2**	**18.0**	**18.0**	**18.3**	**18.4**	**18.5**	**18.7**
Central GDP, Low Prices (CL)								
Solid Fuel	3.0	1.7	1.6	1.2	0.9	0.6	0.5	0.4
Oil	1.0	1.0	1.1	0.6	0.3	0.3	0.2	0.2
Gas	9.7	10.2	11.9	12.7	13.4	13.8	14.0	14.1
Electricity	3.0	3.2	3.4	3.5	3.7	3.7	3.9	4.1
Total	**16.7**	**16.2**	**18.0**	**18.0**	**18.3**	**18.5**	**18.6**	**18.8**
High GDP, Low Prices (HL)								
Solid Fuel	3.0	1.7	1.6	1.2	0.8	0.6	0.4	0.3
Oil	1.0	1.0	1.1	0.6	0.3	0.3	0.2	0.2
Gas	9.7	10.2	11.9	12.7	13.5	13.8	14.0	14.1
Electricity	3.0	3.2	3.4	3.5	3.7	3.8	4.0	4.2
Total	**16.7**	**16.2**	**18.0**	**18.1**	**18.4**	**18.5**	**18.7**	**18.8**

1. Totals may not equal sum of parts due to rounding.

TABLE A.2 - HIGH FUEL PRICES: DOMESTIC SECTOR - FINAL ENERGY DEMAND BY FUEL

Billion Therms	1985	1990	1995	2000	2005	2010	2015	2020
Low GDP, High Prices (LH)								
Solid Fuel	3.0	1.7	1.6	1.2	0.9	0.7	0.5	0.5
Oil	1.0	1.0	0.9	0.4	0.2	0.1	0.1	0.0
Gas	9.7	10.2	11.9	12.7	13.3	13.6	13.7	13.7
Electricity	3.0	3.2	3.4	3.4	3.5	3.6	3.7	3.9
Total	16.7	16.2	17.8	17.7	17.9	17.9	18.0	18.1
Central GDP, High Prices (CH)								
Solid Fuel	3.0	1.7	1.6	1.2	0.9	0.6	0.5	0.4
Oil	1.0	1.0	0.9	0.4	0.2	0.1	0.1	0.0
Gas	9.7	10.2	11.9	12.7	13.3	13.6	13.7	13.7
Electricity	3.0	3.2	3.4	3.4	3.6	3.6	3.9	4.1
Total	16.7	16.2	17.8	17.7	17.9	18.0	18.1	18.2
High GDP, High Prices (HH)								
Solid Fuel	3.0	1.7	1.6	1.2	0.8	0.6	0.4	0.3
Oil	1.0	1.0	0.9	0.4	0.2	0.1	0.1	0.0
Gas	9.7	10.2	11.9	12.7	13.3	13.6	13.7	13.7
Electricity	3.0	3.2	3.4	3.5	3.6	3.7	4.0	4.2
Total	16.7	16.2	17.8	17.7	18.0	18.0	18.2	18.3

1. Totals may not equal sum of parts due to rounding.

TABLE A.3 - LOW FUEL PRICES: IRON AND STEEL SECTOR - FINAL ENERGY DEMAND BY FUEL

Billion Therms	1985	1990	1995	2000	2005	2010	2015	2020
Low GDP, Low Prices (LL)								
Solid Fuel	1.8	1.7	1.5	1.6	1.6	1.5	1.4	1.4
Oil	0.3	0.3	0.3	0.3	0.3	0.4	0.4	0.4
Gas	0.7	0.7	0.6	0.7	0.8	0.8	0.9	0.9
Electricity	0.3	0.3	0.3	0.3	0.4	0.4	0.4	0.4
Total	**3.0**	**3.0**	**2.7**	**2.9**	**3.1**	**3.1**	**3.1**	**3.0**
Central GDP, Low Prices (CL)								
Solid Fuel	1.8	1.7	1.5	1.8	1.9	1.7	1.6	1.6
Oil	0.3	0.3	0.3	0.4	0.4	0.4	0.4	0.4
Gas	0.7	0.7	0.6	0.8	0.9	1.1	1.1	1.1
Electricity	0.3	0.3	0.3	0.4	0.4	0.5	0.5	0.5
Total	**3.0**	**3.0**	**2.8**	**3.3**	**3.6**	**3.7**	**3.7**	**3.6**
High GDP, Low Prices (HL)								
Solid Fuel	1.8	1.7	1.6	1.8	1.9	1.7	1.6	1.6
Oil	0.3	0.3	0.3	0.4	0.4	0.4	0.4	0.4
Gas	0.7	0.7	0.6	0.8	0.9	1.1	1.2	1.2
Electricity	0.3	0.3	0.3	0.4	0.4	0.5	0.5	0.5
Total	**3.0**	**3.0**	**3.0**	**3.4**	**3.7**	**3.8**	**3.8**	**3.7**

1. Gas excludes Blast Furnace Gas.
2. Electricity Projections are for Total Electricity: Public Supply and Own Generation.
3. Fossil Fuels exclude inputs to own generation.
4. Totals may not equal sum of parts due to rounding.

TABLE A.4 - HIGH FUEL PRICES: IRON AND STEEL SECTOR - FINAL ENERGY DEMAND BY FUEL

Billion Therms	1985	1990	1995	2000	2005	2010	2015	2020
Low GDP, High Prices (LH)								
Solid Fuel	1.8	1.7	1.5	1.6	1.6	1.5	1.4	1.3
Oil	0.3	0.3	0.3	0.3	0.3	0.3	0.3	0.2
Gas	0.7	0.7	0.6	0.7	0.8	0.9	1.0	1.0
Electricity	0.3	0.3	0.3	0.4	0.4	0.4	0.4	0.4
Total	**3.0**	**3.0**	**2.7**	**2.9**	**3.2**	**3.2**	**3.1**	**3.0**
Central GDP, High Prices (CH)								
Solid Fuel	1.8	1.7	1.5	1.8	1.9	1.7	1.6	1.6
Oil	0.3	0.3	0.3	0.3	0.4	0.3	0.3	0.3
Gas	0.7	0.7	0.6	0.8	1.0	1.2	1.2	1.2
Electricity	0.3	0.3	0.3	0.4	0.4	0.5	0.5	0.5
Total	**3.0**	**3.0**	**2.8**	**3.3**	**3.6**	**3.7**	**3.7**	**3.5**
High GDP, High Prices (HH)								
Solid Fuel	1.8	1.7	1.6	1.8	1.9	1.7	1.6	1.6
Oil	0.3	0.3	0.3	0.3	0.4	0.3	0.3	0.3
Gas	0.7	0.7	0.6	0.8	1.0	1.2	1.3	1.2
Electricity	0.3	0.3	0.3	0.4	0.4	0.5	0.5	0.5
Total	**3.0**	**3.0**	**2.9**	**3.4**	**3.7**	**3.8**	**3.8**	**3.6**

1. Gas excludes Blast Furnace Gas.
2. Electricity Projections are for Total Electricity: Public Supply and Own Generation.
3. Fossil Fuels exclude inputs to own generation.
4. Totals may not equal sum of parts due to rounding.

TABLE A.5 - LOW FUEL PRICES: OTHER INDUSTRY SECTORS - FINAL ENERGY DEMAND BY FUEL

Billion Therms	1985	1990	1995	2000	2005	2010	2015	2020
Low GDP, Low Prices (LL)								
Solid Fuel	2.0	1.7	1.4	1.2	1.1	0.9	0.8	0.8
Oil	3.6	2.9	3.0	2.7	2.5	2.4	2.3	2.3
Gas	5.5	4.3	4.1	4.7	5.1	5.3	5.4	5.5
Electricity	2.4	3.1	3.2	3.8	4.2	4.5	4.7	5.1
Total	**13.5**	**12.2**	**11.7**	**12.4**	**12.9**	**13.1**	**13.3**	**13.7**
Central GDP, Low Prices (CL)								
Solid Fuel	2.0	1.7	1.4	1.3	1.1	0.9	0.7	0.7
Oil	3.6	2.9	3.0	2.8	2.7	2.4	2.3	2.3
Gas	5.5	4.3	4.1	4.8	5.4	5.8	6.2	6.6
Electricity	2.4	3.1	3.2	4.0	4.5	4.9	5.3	5.9
Total	**13.5**	**12.2**	**11.8**	**12.9**	**13.8**	**14.0**	**14.6**	**15.4**
High GDP, Low Prices (HL)								
Solid Fuel	2.0	1.7	1.4	1.3	1.2	0.9	0.7	0.6
Oil	3.6	2.9	3.0	2.9	2.8	2.4	2.3	2.3
Gas	5.5	4.3	4.2	4.9	5.6	6.2	6.8	7.2
Electricity	2.4	3.1	3.3	4.1	4.7	5.2	5.8	6.4
Total	**13.5**	**12.2**	**11.9**	**13.2**	**14.3**	**14.7**	**15.5**	**16.5**

1. Other Industry includes Construction.
2. Gas excludes non-energy gas.
3. Electricity projections are for Total Electricity: Public Supply and Own Generation.
4. Fossil Fuels exclude inputs into Own Generation.
5. Totals may not equal sum of parts due to rounding.

TABLE A.6 - HIGH FUEL PRICES: OTHER INDUSTRY SECTORS - FINAL ENERGY DEMAND BY FUEL

Billion Therms	1985	1990	1995	2000	2005	2010	2015	2020
Low GDP, High Prices (LH)								
Solid Fuel	2.0	1.7	1.4	1.2	1.1	1.2	1.5	2.1
Oil	3.6	2.9	2.8	2.2	1.9	1.6	1.4	1.3
Gas	5.5	4.3	4.1	4.6	5.0	5.2	5.4	5.6
Electricity	2.4	3.1	3.2	3.6	4.1	4.2	4.6	5.0
Total	13.5	12.2	11.4	11.6	12.0	12.1	13.0	14.0
Central GDP, High Prices (CH)								
Solid Fuel	2.0	1.7	1.4	1.2	1.2	1.2	1.6	2.3
Oil	3.6	2.9	2.8	2.2	2.0	1.7	1.5	1.4
Gas	5.5	4.4	4.1	4.8	5.3	5.6	6.1	6.4
Electricity	2.4	3.1	3.2	3.8	4.3	4.6	5.2	5.8
Total	13.5	12.2	11.6	12.1	12.9	13.2	14.4	15.9
High GDP, High Prices (HH)								
Solid Fuel	2.0	1.7	1.4	1.3	1.2	1.2	1.6	2.3
Oil	3.6	2.9	2.8	2.3	2.1	1.8	1.6	1.4
Gas	5.5	4.3	4.2	4.9	5.5	6.0	6.6	7.1
Electricity	2.4	3.1	3.3	3.9	4.5	4.9	5.7	6.4
Total	13.5	12.2	11.7	12.4	13.4	13.9	15.4	17.2

1. Other Industry includes Construction.
2. Gas excludes non-energy gas.
3. Electricity projections are for Total Electricity: Public Supply and Own Generation.
4. Fossil Fuels exclude inputs into Own Generation.
5. Totals may not equal sum of parts due to rounding.

TABLE A.7 - LOW FUEL PRICES: SERVICE SECTOR - FINAL ENERGY DEMAND BY FUEL

Billion Therms	1985	1990	1995	2000	2005	2010	2015	2020
Low GDP, Low Prices (LL)								
Solid Fuel	0.6	0.5	0.3	0.2	0.2	0.2	0.2	0.2
Oil	2.4	1.8	1.7	1.3	1.1	1.0	1.0	1.0
Gas	2.7	3.0	3.2	3.8	4.3	4.7	5.0	5.3
Electricity	2.2	2.6	3.0	3.3	3.5	3.6	3.7	3.8
Total	**7.8**	**7.8**	**8.1**	**8.7**	**9.1**	**9.5**	**9.9**	**10.4**
Central GDP, Low Prices (CL)								
Solid Fuel	0.6	0.5	0.3	0.2	0.2	0.2	0.2	0.2
Oil	2.4	1.8	1.7	1.4	1.2	1.1	1.0	1.0
Gas	2.7	3.0	3.2	3.9	4.4	4.9	5.4	5.7
Electricity	2.2	2.6	3.0	3.4	3.6	3.7	3.9	4.1
Total	**7.8**	**7.8**	**8.2**	**8.8**	**9.4**	**9.9**	**10.5**	**11.0**
High GDP, Low Prices (HL)								
Solid Fuel	0.6	0.5	0.3	0.2	0.2	0.2	0.2	0.2
Oil	2.4	1.8	1.7	1.4	1.2	1.1	1.0	1.0
Gas	2.7	3.0	3.3	3.9	4.5	5.1	5.6	6.0
Electricity	2.2	2.6	3.0	3.4	3.6	3.8	4.0	4.2
Total	**7.8**	**7.8**	**8.2**	**8.9**	**9.6**	**10.2**	**10.9**	**11.5**

1. Service Sector includes Commercial, Public Administration and Agriculture sectors.
2. Gas excludes non-energy gas.
3. Electricity projections are for Total Electricity: Public Supply and Own Generation.
4. Fossil Fuels exclude inputs into Own Generation.
5. Totals may not equal sum of parts due to rounding.

TABLE A.8 - HIGH FUEL PRICES: SERVICE SECTOR - FINAL ENERGY DEMAND BY FUEL

Billion Therms	1985	1990	1995	2000	2005	2010	2015	2020
Low GDP, High Prices (LH)								
Solid Fuel	0.6	0.5	0.3	0.2	0.2	0.2	0.2	0.2
Oil	2.4	1.8	1.6	1.0	0.7	0.6	0.5	0.5
Gas	2.7	3.0	3.3	4.2	4.8	5.3	5.7	6.0
Electricity	2.2	2.6	3.0	3.2	3.4	3.5	3.6	3.8
Total	7.8	7.8	8.1	8.7	9.2	9.6	10.0	10.4
Central GDP, High Prices (CH)								
Solid Fuel	0.6	0.5	0.3	0.3	0.2	0.2	0.2	0.2
Oil	2.4	1.8	1.6	1.0	0.7	0.6	0.5	0.5
Gas	2.7	3.0	3.3	4.3	5.0	5.6	6.0	6.4
Electricity	2.2	2.6	3.0	3.3	3.5	3.6	3.8	4.0
Total	7.8	7.8	8.2	8.8	9.4	10.0	10.6	11.1
High GDP, High Prices (HH)								
Solid Fuel	0.6	0.5	0.3	0.3	0.2	0.2	0.2	0.2
Oil	2.4	1.8	1.6	1.0	0.7	0.6	0.5	0.5
Gas	2.7	3.0	3.4	4.3	5.1	5.8	6.3	6.7
Electricity	2.2	2.6	3.0	3.3	3.6	3.7	4.0	4.2
Total	7.8	7.8	8.2	8.9	9.7	10.3	11.0	11.6

1. Service Sector includes Commercial, Public Administration and Agriculture sectors.
2. Gas excludes non-energy gas.
3. Electricity projections are for Total Electricity: Public Supply and Own Generation.
4. Fossil Fuels exclude inputs into Own Generation.
5. Totals may not equal sum of parts due to rounding.

TABLE A.9 - LOW FUEL PRICES: TRANSPORT SECTOR - FINAL ENERGY DEMAND BY FUEL

Billion Therms	1985	1990	1995	2000	2005	2010	2015	2020
Low GDP, Low Prices (LL)								
Motor Spirit	9.2	10.8	11.1	10.9	10.4	10.5	11.0	11.7
Diesel	3.1	4.6	5.5	6.7	8.1	9.1	10.1	11.1
Other	3.1	3.8	4.4	4.8	5.3	6.2	7.4	8.8
Total	15.4	19.3	20.9	22.3	23.8	25.8	28.5	31.6
Central GDP, Low Prices (CL)								
Motor Spirit	9.2	10.8	11.1	11.0	10.6	10.7	11.4	12.2
Diesel	3.1	4.6	5.5	6.8	8.5	9.8	11.1	12.6
Other	3.1	3.8	4.4	5.0	5.9	6.9	8.2	9.8
Total	15.4	19.3	21.0	22.8	24.9	27.4	30.7	34.6
High GDP, Low Prices (HL)								
Motor Spirit	9.2	10.8	11.1	11.1	10.7	11.0	11.7	12.6
Diesel	3.1	4.6	5.5	7.0	8.9	10.4	12.1	14.0
Other	3.1	3.8	4.5	5.3	6.5	7.6	9.1	11.0
Total	15.4	19.3	21.1	23.4	26.1	29.1	33.0	37.5

1. 'Other' includes Aviation Fuel, Gasoil for Water and Rail Transport, Heavy Fuel Oil and Electricity.
2. Totals may not equal sum of parts due to rounding.

TABLE A.10 - HIGH FUEL PRICES: TRANSPORT SECTOR - FINAL ENERGY DEMAND BY FUEL

Billion Therms	1985	1990	1995	2000	2005	2010	2015	2020
Low GDP, High Prices (LH)								
Motor Spirit	9.2	10.8	11.0	10.5	9.7	9.4	9.6	9.9
Diesel	3.1	4.6	5.5	6.6	7.9	8.8	9.7	10.6
Other	3.1	3.8	4.3	4.6	5.1	5.9	7.0	8.3
Total	15.4	19.3	20.7	21.7	22.6	24.1	26.2	28.8
Central GDP, High Prices (CH)								
Motor Spirit	9.2	10.8	11.0	10.6	9.9	9.6	10.0	10.4
Diesel	3.1	4.6	5.5	6.7	8.3	9.5	10.7	12.0
Other	3.1	3.8	4.3	4.8	5.6	6.5	7.7	9.3
Total	15.4	19.3	20.8	22.2	23.7	25.7	28.4	31.7
High GDP, High Prices (HH)								
Motor Spirit	9.2	10.8	11.0	10.7	10.0	9.9	10.3	10.8
Diesel	3.1	4.6	5.5	6.9	8.7	10.1	11.7	13.5
Other	3.1	3.8	4.4	5.1	6.2	7.3	8.6	10.3
Total	15.4	19.3	20.9	22.7	24.9	27.3	30.6	34.6

1. 'Other' includes Aviation Fuel, Gasoil for Water and Rail Transport, Heavy Fuel Oil and Electricity.
2. Totals may not equal sum of parts due to rounding.

TABLE A.11 - LOW FUEL PRICES: TOTAL FINAL USERS - FINAL ENERGY DEMAND BY FUEL

Billion Therms	1985	1990	1995	2000	2005	2010	2015	2020
Low GDP, Low Prices (LL)								
Solid Fuel	7.4	5.6	4.8	4.2	3.8	3.3	3.0	2.8
Oil	22.4	25.1	26.8	27.1	28.0	29.6	32.2	35.2
Gas	18.6	18.3	19.8	21.9	23.6	24.6	25.3	25.9
Electricity	8.0	9.4	10.0	11.1	11.9	12.4	12.8	13.4
Total	**56.3**	**58.4**	**61.4**	**64.3**	**67.3**	**70.0**	**73.3**	**77.4**
Central GDP, Low Prices (CL)								
Solid Fuel	7.4	5.6	4.8	4.5	4.1	3.5	3.1	2.8
Oil	22.4	25.1	26.9	27.8	29.4	31.3	34.4	38.2
Gas	18.6	18.3	19.9	22.2	24.2	25.6	26.7	27.6
Electricity	8.0	9.4	10.1	11.4	12.4	13.0	13.8	14.7
Total	**56.3**	**58.4**	**61.7**	**65.9**	**70.0**	**73.5**	**78.0**	**83.3**
High GDP, Low Prices (HL)								
Solid Fuel	7.4	5.6	5.0	4.5	4.1	3.4	3.0	2.7
Oil	22.4	25.1	27.1	28.5	30.6	33.0	36.7	41.2
Gas	18.6	18.3	20.0	22.4	24.6	26.3	27.6	28.6
Electricity	8.0	9.4	10.1	11.5	12.7	13.5	14.5	15.6
Total	**56.3**	**58.4**	**62.2**	**66.9**	**72.0**	**76.3**	**81.8**	**88.2**

1. Totals may not equal sum of parts due to rounding.

TABLE A.12 - HIGH FUEL PRICES: TOTAL FINAL USERS - FINAL ENERGY DEMAND BY FUEL

Billion Therms	1985	1990	1995	2000	2005	2010	2015	2020
Low GDP, High Prices (LH)								
Solid Fuel	7.4	5.6	4.7	4.2	3.8	3.5	3.7	4.1
Oil	22.4	25.1	26.1	25.3	25.5	26.4	28.3	30.6
Gas	18.6	18.3	19.9	22.2	24.0	25.0	25.8	26.2
Electricity	8.0	9.4	10.0	10.8	11.5	11.9	12.6	13.3
Total	**56.3**	**58.4**	**60.7**	**62.5**	**64.9**	**66.9**	**70.4**	**74.3**
Central GDP, High Prices (CH)								
Solid Fuel	7.4	5.6	4.8	4.4	4.1	3.8	3.9	4.4
Oil	22.4	25.1	26.2	26.0	26.8	28.2	30.7	33.7
Gas	18.6	18.3	20.0	22.6	24.6	26.0	27.0	27.7
Electricity	8.0	9.4	10.1	11.1	12.0	12.6	13.6	14.6
Total	**56.3**	**58.4**	**61.1**	**64.1**	**67.6**	**70.6**	**75.2**	**80.4**
High GDP, High Prices (HH)								
Solid Fuel	7.4	5.6	4.9	4.5	4.2	3.8	3.8	4.4
Oil	22.4	25.1	26.4	26.6	28.1	29.9	33.0	36.7
Gas	18.6	18.3	20.1	22.8	25.0	26.6	27.9	28.7
Electricity	8.0	9.4	10.2	11.2	12.3	13.1	14.3	15.5
Total	**56.3**	**58.4**	**61.6**	**65.1**	**69.6**	**73.4**	**79.0**	**85.4**

1. Totals may not equal sum of parts due to rounding.

ANNEX B

PROJECTIONS BY END USE

DOMESTIC SECTOR
- Table B.1 - Final Energy Demand by End Use - Low GDP growth
- Table B.2 - Final Energy Demand by End Use - Central GDP growth
- Table B.3 - Final Energy Demand by End Use - High GDP growth

SERVICE SECTOR (Not Including Agriculture)
- Table B.1 - Final Energy Demand by End Use - Low GDP growth
- Table B.2 - Final Energy Demand by End Use - Central GDP growth
- Table B.3 - Final Energy Demand by End Use - High GDP growth

NOTES:

- End use data are far less reliable than the aggregate data shown in the Digest of UK Energy Statistics (DUKES), as end use data rely on infrequent surveys.

- The domestic sector breakdown of aggregate data into its end uses is based on information supplied by the Building Research Establishment (BRE).

- Service sector breakdown by end use is derived using:
  BRE Information Paper IP 16/94 (October 1994);
  Energy Efficiency Office (EEO) - "Energy use and energy efficiency in UK commercial and public buildings up to year 2000", HMSO (1988);
  Electricity Association (EA) - "UK Electricity" (1991, 1992, 1993).

- Although assistance provided by the BRE and the above publications was helpful, ultimate responsibility for the end use data actually used for EP65 remains with the DTI.

TABLE B.1 - LOW GDP: DOMESTIC SECTOR - FINAL ENERGY DEMAND BY END USE

Billion Therms	1990	1995	2000	2005	2010	2015	2020
Low GDP, Low Prices (LL)							
Space & Water Heating - Solid Fuel	1.8	1.6	1.2	0.9	0.7	0.6	0.5
Space & Water Heating - Oil	1.0	1.1	0.6	0.3	0.2	0.2	0.2
Space & Water Heating - Gas	9.7	11.4	12.2	12.9	13.3	13.4	13.5
Space & Water Heating - Electricity	0.9	1.2	1.3	1.5	1.5	1.5	1.6
Cooking - Gas	0.5	0.5	0.5	0.5	0.5	0.5	0.5
Cooking - Electricity	0.2	0.2	0.2	0.2	0.2	0.2	0.2
Appliances - Electricity	2.0	2.0	2.0	2.0	2.0	2.1	2.1
Total	16.2	18.0	18.0	18.3	18.4	18.5	18.6
Low GDP, High Prices (LH)							
Space & Water Heating - Solid Fuel	1.8	1.6	1.2	0.9	0.7	0.5	0.5
Space & Water Heating - Oil	1.0	0.9	0.4	0.2	0.1	0.1	0.0
Space & Water Heating - Gas	9.7	11.4	12.2	12.8	13.1	13.1	13.2
Space & Water Heating - Electricity	0.9	1.2	1.2	1.4	1.4	1.5	1.6
Cooking - Gas	0.5	0.5	0.5	0.5	0.5	0.5	0.5
Cooking - Electricity	0.2	0.2	0.2	0.2	0.2	0.2	0.2
Appliances - Electricity	2.0	2.0	2.0	2.0	2.0	2.1	2.1
Total	16.2	17.8	17.7	17.9	17.9	18.0	18.1

1. Totals may not equal sum of parts due to rounding.

TABLE B.2 - CENTRAL GDP: DOMESTIC SECTOR - FINAL ENERGY DEMAND BY END USE

Billion Therms	1990	1995	2000	2005	2010	2015	2020
Central GDP, Low Prices (CL)							
Space & Water Heating - Solid Fuel	1.8	1.6	1.2	0.9	0.6	0.5	0.4
Space & Water Heating - Oil	1.0	1.1	0.6	0.3	0.3	0.2	0.2
Space & Water Heating - Gas	9.7	11.4	12.2	12.9	13.3	13.5	13.6
Space & Water Heating - Electricity	0.9	1.2	1.3	1.5	1.5	1.6	1.7
Cooking - Gas	0.5	0.5	0.5	0.5	0.5	0.5	0.5
Cooking - Electricity	0.2	0.2	0.2	0.2	0.2	0.2	0.2
Appliances - Electricity	2.0	2.0	2.0	2.0	2.0	2.1	2.2
Total	16.2	18.0	18.0	18.3	18.5	18.6	18.8
Central GDP, High Prices (CH)							
Space & Water Heating - Solid Fuel	1.8	1.6	1.2	0.9	0.6	0.5	0.4
Space & Water Heating - Oil	1.0	0.9	0.4	0.2	0.1	0.1	0.0
Space & Water Heating - Gas	9.7	11.4	12.2	12.8	13.1	13.2	13.2
Space & Water Heating - Electricity	0.9	1.2	1.3	1.4	1.4	1.6	1.7
Cooking - Gas	0.5	0.5	0.5	0.5	0.5	0.5	0.5
Cooking - Electricity	0.2	0.2	0.2	0.2	0.2	0.2	0.2
Appliances - Electricity	2.0	2.0	2.0	2.0	2.0	2.1	2.2
Total	16.2	17.8	17.7	17.9	18.0	18.1	18.2

1. Totals may not equal sum of parts due to rounding.

TABLE B.3 - HIGH GDP: DOMESTIC SECTOR - FINAL ENERGY DEMAND BY END USE

Billion Therms	1990	1995	2000	2005	2010	2015	2020
High GDP, Low Prices (HL)							
Space & Water Heating - Solid Fuel	1.8	1.6	1.2	0.8	0.6	0.4	0.3
Space & Water Heating - Oil	1.0	1.1	0.6	0.3	0.3	0.2	0.2
Space & Water Heating - Gas	9.7	11.4	12.2	13.0	13.3	13.5	13.6
Space & Water Heating - Electricity	0.9	1.2	1.4	1.5	1.6	1.7	1.8
Cooking - Gas	0.5	0.5	0.5	0.5	0.5	0.5	0.5
Cooking - Electricity	0.2	0.2	0.2	0.2	0.2	0.2	0.2
Appliances - Electricity	2.0	2.0	2.0	2.0	2.0	2.1	2.2
Total	16.2	18.0	18.1	18.4	18.5	18.7	18.8
High GDP, High Prices (HH)							
Space & Water Heating - Solid Fuel	1.8	1.6	1.2	0.8	0.6	0.4	0.3
Space & Water Heating - Oil	1.0	0.9	0.4	0.2	0.1	0.1	0.0
Space & Water Heating - Gas	9.7	11.4	12.2	12.8	13.1	13.2	13.2
Space & Water Heating - Electricity	0.9	1.2	1.3	1.4	1.5	1.7	1.8
Cooking - Gas	0.5	0.5	0.5	0.5	0.5	0.5	0.5
Cooking - Electricity	0.2	0.2	0.2	0.2	0.2	0.2	0.2
Appliances - Electricity	2.0	2.0	2.0	2.0	2.0	2.1	2.2
Total	16.2	17.8	17.7	18.0	18.0	18.2	18.3

1. Totals may not equal sum of parts due to rounding.

TABLE B4 - LOW GDP: SERVICE SECTOR - FINAL ENERGY DEMAND BY END USE

Billion Therms	1990	1995	2000	2005	2010	2015	2020
Low GDP, Low Prices (LL)							
Space & Water Heating - Solid Fuel	0.4	0.2	0.1	0.1	0.1	0.1	0.1
Space & Water Heating - Oil	1.4	1.5	1.1	0.9	0.8	0.8	0.8
Space & Water Heating - Gas	2.6	3.0	3.5	3.8	4.1	4.5	4.7
Space & Water Heating - Electricity	0.5	0.5	0.6	0.6	0.6	0.6	0.6
Cooking - Gas	0.3	0.3	0.3	0.3	0.3	0.4	0.4
Cooking - Electricity	0.2	0.2	0.3	0.3	0.3	0.3	0.3
Lighting - Electricity	0.8	0.8	0.9	0.9	1.0	1.0	1.1
Air Conditioning - Electricity	0.3	0.3	0.4	0.4	0.5	0.6	0.7
Other End Uses - Electricity	0.7	1.0	1.0	1.0	1.0	1.0	0.9
Total	**7.2**	**7.8**	**8.2**	**8.5**	**8.9**	**9.3**	**9.7**
Low GDP, High Prices (LH)							
Space & Water Heating - Solid Fuel	0.4	0.2	0.1	0.1	0.1	0.1	0.1
Space & Water Heating - Oil	1.4	1.3	0.7	0.5	0.4	0.3	0.3
Space & Water Heating - Gas	2.6	3.0	3.7	4.2	4.6	4.9	5.2
Space & Water Heating - Electricity	0.5	0.5	0.5	0.6	0.6	0.6	0.6
Cooking - Gas	0.3	0.3	0.3	0.3	0.3	0.3	0.3
Cooking - Electricity	0.2	0.2	0.2	0.3	0.3	0.3	0.4
Lighting - Electricity	0.8	0.8	0.9	0.9	1.0	1.0	1.1
Air Conditioning - Electricity	0.3	0.3	0.4	0.4	0.5	0.6	0.7
Other End Uses - Electricity	0.7	1.0	1.0	1.0	1.0	1.0	0.9
Total	**7.2**	**7.7**	**8.1**	**8.4**	**8.8**	**9.2**	**9.5**

1. Service sector end uses data excludes the Agriculture sector
2. Totals may not equal sum of parts due to rounding.

TABLE B.5 - CENTRAL GDP: SERVICE SECTOR - FINAL ENERGY DEMAND BY END USE

Billion Therms	1990	1995	2000	2005	2010	2015	2020
Central GDP, Low Prices (CL)							
Space & Water Heating - Solid Fuel	0.4	0.2	0.1	0.1	0.1	0.1	0.1
Space & Water Heating - Oil	1.4	1.5	1.1	0.9	0.8	0.8	0.8
Space & Water Heating - Gas	2.6	3.0	3.5	4.0	4.4	4.8	5.1
Space & Water Heating - Electricity	0.5	0.5	0.6	0.6	0.6	0.6	0.6
Cooking - Gas	0.3	0.3	0.4	0.4	0.4	0.4	0.4
Cooking - Electricity	0.2	0.2	0.3	0.3	0.3	0.3	0.4
Lighting - Electricity	0.8	0.8	0.9	1.0	1.0	1.1	1.2
Air Conditioning - Electricity	0.3	0.3	0.4	0.5	0.6	0.6	0.7
Other End Uses - Electricity	0.7	1.0	1.1	1.1	1.1	1.0	1.0
Total	**7.2**	**7.8**	**8.3**	**8.8**	**9.3**	**9.8**	**10.3**
Central GDP, High Prices (CH)							
Space & Water Heating - Solid Fuel	0.4	0.2	0.2	0.1	0.1	0.1	0.1
Space & Water Heating - Oil	1.4	1.3	0.8	0.5	0.4	0.3	0.3
Space & Water Heating - Gas	2.6	3.1	3.8	4.4	4.8	5.2	5.5
Space & Water Heating - Electricity	0.5	0.5	0.5	0.6	0.6	0.6	0.6
Cooking - Gas	0.3	0.3	0.4	0.3	0.4	0.4	0.3
Cooking - Electricity	0.2	0.2	0.3	0.3	0.3	0.3	0.4
Lighting - Electricity	0.8	0.8	0.9	1.0	1.0	1.1	1.1
Air Conditioning - Electricity	0.3	0.3	0.4	0.5	0.6	0.6	0.7
Other End Uses - Electricity	0.7	1.0	1.1	1.1	1.1	1.0	1.0
Total	**7.2**	**7.8**	**8.2**	**8.7**	**9.2**	**9.7**	**10.2**

1. Service sector end uses data excludes the Agriculture sector
2. Totals may not equal sum of parts due to rounding.

TABLE B.6 - HIGH GDP: SERVICE SECTOR - FINAL ENERGY DEMAND BY END USE

Billion Therms	1990	1995	2000	2005	2010	2015	2020
High GDP, Low Prices (HL)							
Space & Water Heating - Solid Fuel	0.4	0.2	0.1	0.1	0.1	0.1	0.1
Space & Water Heating - Oil	1.4	1.5	1.1	0.9	0.8	0.8	0.8
Space & Water Heating - Gas	2.6	3.0	3.6	4.1	4.6	5.0	5.4
Space & Water Heating - Electricity	0.5	0.5	0.6	0.6	0.6	0.6	0.6
Cooking - Gas	0.3	0.3	0.4	0.4	0.4	0.4	0.4
Cooking - Electricity	0.2	0.2	0.3	0.3	0.3	0.4	0.4
Lighting - Electricity	0.8	0.8	0.9	1.0	1.1	1.1	1.2
Air Conditioning - Electricity	0.3	0.3	0.4	0.5	0.6	0.7	0.8
Other End Uses - Electricity	0.7	1.0	1.1	1.1	1.1	1.1	1.1
Total	7.2	7.9	8.4	9.0	9.6	10.2	10.8
High GDP, High Prices (HH)							
Space & Water Heating - Solid Fuel	0.4	0.2	0.2	0.1	0.1	0.1	0.1
Space & Water Heating - Oil	1.4	1.3	0.8	0.5	0.4	0.3	0.3
Space & Water Heating - Gas	2.6	3.1	3.9	4.5	5.0	5.4	5.8
Space & Water Heating - Electricity	0.5	0.5	0.5	0.6	0.6	0.6	0.6
Cooking - Gas	0.3	0.3	0.4	0.4	0.4	0.4	0.4
Cooking - Electricity	0.2	0.2	0.3	0.3	0.3	0.4	0.4
Lighting - Electricity	0.8	0.8	0.9	1.0	1.0	1.1	1.2
Air Conditioning - Electricity	0.3	0.3	0.4	0.5	0.6	0.7	0.8
Other End Uses - Electricity	0.7	1.0	1.1	1.1	1.1	1.1	1.1
Total	7.2	7.8	8.3	8.9	9.5	10.1	10.6

1. Service sector end uses data excludes the Agriculture sector
2. Totals may not equal sum of parts due to rounding.

ANNEX C

Table C1: Total ESI Capacity, GW *

EP65	1990	1995	2000	2005	2010	2015	2020
LL	70.2	69.0	74.8	77.4	76.6	78.0	80.5
LH	70.2	69.0	74.8	75.5	73.2	76.7	79.9
CL	70.2	69.0	76.4	80.1	80.5	84.1	88.7
CH	70.2	69.0	74.8	78.1	77.7	82.9	87.7
HL	70.2	69.0	77.4	82.0	83.8	88.7	94.1
HH	70.2	69.0	75.7	80.0	81.0	87.5	93.7

EP59	1990	1995	2000	2005	2010	2015	2020
LL	70.2	73.1	76.5	82.2	87.8	95.2	101.2
LH	70.2	73.1	75.6	81.0	86.3	93.2	100.1
CL	70.2	73.4	78.8	85.6	94.1	100.4	114.8
CH	70.2	73.3	77.6	84.1	91.9	101.0	112.7
HL	70.2	73.5	82.5	95.7	111.1	132.1	159.5
HH	70.2	73.1	80.9	93.5	108.8	130.0	148.3

*includes 'Major Power Producers': those companies whose main business is power generation

Table C2: ESI Plant Capacity, GW

LL scenario

	coal	coal FGD	oil	oil FGD	IGCC	CCGT	nuclear	renewables
1990	40	0	12	0	0	0	11	1
1995	29	4	8	0	0	9	13	2
2000	17	10	8	0	0	16	12	3
2005	13	10	6	2	0	17	9	3
2010	6	6	6	2	0	28	7	3
2015	2	6	3	2	3	29	3	4
2020	2	1	3	0	12	31	1	4

CL scenario

	coal	coal FGD	oil	oil FGD	IGCC	CCGT	nuclear	renewables
1990	40	0	12	0	0	0	11	1
1995	29	4	8	0	0	9	13	2
2000	17	10	8	0	0	17	12	3
2005	13	10	6	2	0	19	9	3
2010	6	6	6	2	0	30	7	4
2015	2	6	3	2	5	32	3	4
2020	2	1	3	0	16	33	1	4

HL scenario

	coal	coal FGD	oil	oil FGD	IGCC	CCGT	nuclear	renewables
1990	40	0	12	0	0	0	11	1
1995	29	4	8	0	0	9	13	2
2000	17	10	8	0	0	18	12	3
2005	13	10	6	2	0	21	9	3
2010	6	6	6	2	0	33	7	4
2015	2	6	3	2	7	35	3	4
2020	2	1	3	0	18	37	1	4

Table C2: ESI Plant Capacity, GW (continued)

LH scenario	coal	coal FGD	oil	oil FGD	IGCC	CCGT	nuclear	renewables
1990	40	0	12	0	0	0	11	1
1995	29	4	8	0	0	9	13	2
2000	17	10	8	0	0	16	12	3
2005	13	10	6	2	0	16	9	4
2010	6	6	6	2	3	21	7	4
2015	2	6	3	2	13	22	3	5
2020	2	1	3	0	24	23	1	5

CH scenario	coal	coal FGD	oil	oil FGD	IGCC	CCGT	nuclear	renewables
1990	40	0	12	0	0	0	11	1
1995	29	4	8	0	0	9	13	2
2000	17	10	8	0	0	16	12	3
2005	13	10	6	2	0	16	9	4
2010	6	6	6	2	4	22	7	4
2015	2	6	3	2	16	25	3	5
2020	2	1	3	0	28	26	1	5

HH scenario	coal	coal FGD	oil	oil FGD	IGCC	CCGT	nuclear	renewables
1990	40	0	12	0	0	0	11	1
1995	29	4	8	0	0	9	13	2
2000	17	10	8	0	0	17	12	3
2005	13	10	6	2	0	17	9	4
2010	6	6	6	2	4	25	7	4
2015	2	6	3	2	17	28	3	5
2020	2	1	3	0	30	29	1	5

TABLE C3: ESI FUEL USE*
Low price scenarios

LL	1990	1995	2000	2005	2010	2015	2020
coal	52.0	27.8	14.9	19.5	10.7	13.6	9.4
petroleum	6.8	1.6	6.3	8.1	5.2	7.7	16.1
natural gas	0.0	15.2	24.1	25.8	37.7	39.0	38.1
nuclear	16.3	25.4	25.4	19.7	14.8	7.9	2.7
renewables	0.4	0.5	1.2	1.4	1.5	1.7	1.8
imports	1.0	1.5	1.5	1.5	1.5	1.5	1.5
total	76.5	71.9	73.3	76.0	71.4	71.3	69.5

CL	1990	1995	2000	2005	2010	2015	2020
coal	52.0	28.4	14.6	19.5	10.6	13.1	9.4
petroleum	6.8	1.6	6.4	8.3	6.1	10.3	19.9
natural gas	0.0	15.2	25.7	28.1	40.3	41.8	41.4
nuclear	16.3	25.4	25.4	19.7	14.8	7.9	2.7
renewables	0.4	0.5	1.2	1.5	1.6	1.8	1.9
imports	1.0	1.5	1.5	1.5	1.5	1.5	1.5
total	76.5	72.5	74.8	78.5	74.8	76.3	76.7

HL	1990	1995	2000	2005	2010	2015	2020
coal	52.0	28.9	14.6	19.2	10.7	12.6	8.7
petroleum	6.8	1.6	6.4	8.1	6.2	12.5	22.4
natural gas	0.0	15.2	26.5	30.1	42.7	43.7	43.9
nuclear	16.3	25.4	25.4	19.7	14.8	7.9	2.7
renewables	0.4	0.5	1.2	1.5	1.6	1.8	1.9
imports	1.0	1.5	1.5	1.5	1.5	1.5	1.5
total	76.5	73.1	75.6	80.0	77.4	79.9	81.0

n.b. totals may not equal sum of parts due to rounding

*million tonnes of oil equivalent   n.b. petroleum includes orimulsion

Table C3 continued
High price scenarios

LH	1990	1995	2000	2005	2010	2015	2020
coal	52.0	28.0	21.3	26.8	21.1	29.7	36.8
petroleum	6.8	1.6	0.5	4.1	4.9	4.6	1.1
natural gas	0.0	15.1	22.4	21.3	26.9	25.4	25.1
nuclear	16.3	25.4	25.4	19.7	14.8	7.9	2.7
renewables	0.4	0.5	1.2	1.8	2.0	2.1	2.3
imports	1.0	1.5	1.5	1.5	1.5	1.5	1.5
total	76.5	72.0	72.2	75.0	71.1	71.2	69.4

CH	1990	1995	2000	2005	2010	2015	2020
coal	52.0	28.5	23.3	26.9	21.7	32.2	39.9
petroleum	6.8	1.6	0.6	4.6	5.0	4.6	4.7
natural gas	0.0	15.1	22.4	23.4	29.8	27.9	25.5
nuclear	16.3	25.4	25.4	19.7	14.8	7.9	2.7
renewables	0.4	0.5	1.2	1.8	2.0	2.1	2.3
imports	1.0	1.5	1.5	1.5	1.5	1.5	1.5
total	76.5	72.5	74.3	77.7	74.6	76.3	76.5

HH	1990	1995	2000	2005	2010	2015	2020
coal	52.0	29.3	23.7	27.4	20.6	33.6	39.4
petroleum	6.8	1.6	0.6	4.9	4.2	4.2	7.2
natural gas	0.0	15.1	22.9	24.6	33.9	30.8	28.3
nuclear	16.3	25.4	25.4	19.7	14.8	7.9	2.7
renewables	0.4	0.5	1.2	1.8	2.0	2.2	2.3
imports	1.0	1.5	1.5	1.5	1.5	1.5	1.5
total	76.5	73.3	75.3	79.7	76.9	80.1	81.4

ANNEX D

PRIMARY ENERGY DEMAND

Table D1: EP65 primary energy demand*							
	1990	1995	2000	2005	2010	2015	2020
LL	220.8	222.4	231.4	241.7	245.1	254.1	262.7
LH	220.8	220.7	226.0	234.6	236.8	245.1	253.2
CL	220.8	223.8	236.6	250.7	256.8	269.7	283.1
CH	220.8	222.1	231.7	243.8	248.8	261.1	273.8
HL	220.8	225.5	239.9	257.0	265.9	282.0	298.7
HH	220.8	223.9	235.1	250.4	257.8	273.7	289.8

*million tonnes of oil equivalent

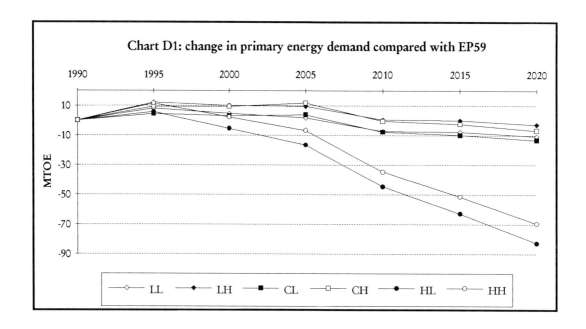

Table D2: Primary demand by fuel*
Low price scenarios

LL	1990	1995	2000	2005	2010	2015	2020
coal	69.5	43.9	29.6	33.2	23.0	25.0	20.0
petroleum	77.8	77.3	83.2	87.6	89.3	99.0	116.1
natural gas	55.8	73.7	90.4	98.1	114.8	118.7	120.0
nuclear&imports	17.3	26.8	26.8	21.1	16.2	9.3	4.1
renewables ◆	0.5	0.7	1.5	1.7	1.8	2.1	2.2
total	220.8	222.4	231.4	241.7	245.1	254.1	262.7

CL	1990	1995	2000	2005	2010	2015	2020
coal	69.5	44.7	30.1	34.2	23.5	24.9	20.5
petroleum	77.8	77.7	85.2	91.6	94.9	107.8	128.1
natural gas	55.8	73.9	93.0	102.1	120.3	125.6	128.2
nuclear&imports	17.3	26.8	26.8	21.1	16.2	9.3	4.1
renewables	0.5	0.7	1.5	1.8	1.9	2.1	2.3
total	220.8	223.8	236.6	250.7	256.8	269.7	283.2

HL	1990	1995	2000	2005	2010	2015	2020
coal	69.5	45.7	30.4	34.0	23.4	24.1	19.5
petroleum	77.8	78.1	86.9	94.9	99.9	116.3	138.4
natural gas	55.8	74.1	94.3	105.3	124.7	130.2	133.8
nuclear&imports	17.3	26.8	26.8	21.1	16.2	9.3	4.1
renewables	0.5	0.7	1.5	1.8	1.9	2.1	2.3
total	220.8	225.5	239.9	257.0	266.1	281.9	298.2

*   million tonnes of oil equivalent

◆   in 1990, solid and gaseous renewables are included in the coal or gas categories. This reflects the current DUKES treatment. This treatment is continued into later years. Thus, 'renewables' relates to hydro and new renewable uses post 1990.

Table D2 continued
High price scenarios

LH	1990	1995	2000	2005	2010	2015	2020
coal	69.5	44.0	35.8	40.5	33.9	42.8	51.0
petroleum	77.8	75.3	72.4	76.7	80.1	85.1	88.3
natural gas	55.8	73.8	89.5	94.3	104.3	105.4	107.0
nuclear&imports	17.3	26.8	26.8	21.1	16.2	9.3	4.1
renewables	0.5	0.7	1.5	2.0	2.3	2.5	2.7
total	220.8	220.7	226.0	234.6	236.8	245.1	253.2

CH	1990	1995	2000	2005	2010	2015	2020
coal	69.5	44.8	38.7	41.7	35.4	46.1	55.2
petroleum	77.8	75.7	74.3	80.8	85.1	91.7	100.4
natural gas	55.8	74.0	90.4	98.2	109.9	111.4	111.3
nuclear&imports	17.3	26.8	26.8	21.1	16.2	9.3	4.1
renewables	0.5	0.7	1.5	2.0	2.3	2.5	2.7
total	220.8	222.1	231.7	243.8	248.8	261.1	273.8

HH	1990	1995	2000	2005	2010	2015	2020
coal	69.5	46.0	39.3	42.3	34.2	47.4	54.6
petroleum	77.8	76.1	76.0	84.5	89.1	97.6	111.1
natural gas	55.8	74.2	91.5	100.5	116.0	116.8	117.2
nuclear&imports	17.3	26.8	26.8	21.1	16.2	9.3	4.1
renewables	0.5	0.7	1.5	2.1	2.3	2.5	2.7
total	220.8	223.9	235.1	250.4	257.8	273.7	289.8

ANNEX E

CARBON DIOXIDE AND SULPHUR DIOXIDE EMISSIONS

	Table E1: UK carbon dioxide emissions, 1990-2020						
	1990	1995	2000	2005	2010	2015	2020
LL	158.3	149.7	146.7	156.8	155.2	164.4	171.4
LH	158.3	148.4	144.1	153.7	153.6	164.6	173.1
CL	158.3	150.8	149.7	162.0	161.8	173.7	184.2
CH	158.3	149.4	148.4	159.2	160.8	175.0	187.8
HL	158.3	152.0	151.5	165.1	166.7	180.8	193.2
HH	158.3	150.9	150.5	163.0	164.9	182.4	197.4

* as million tonnes of carbon

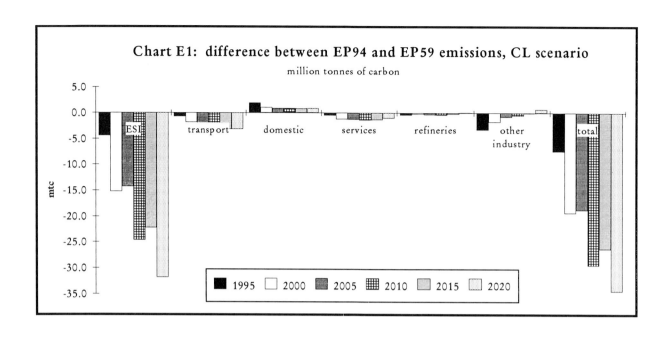
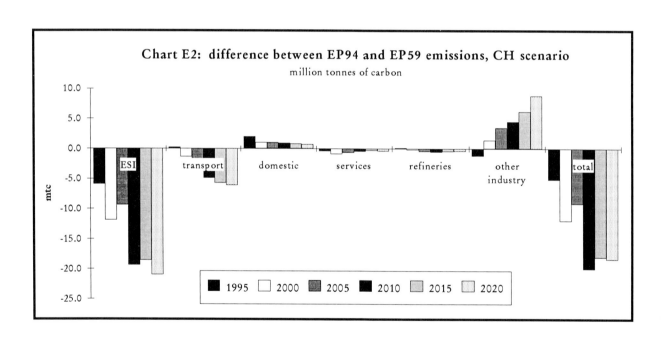

Table E2: UK carbon dioxide emissions by sector*
Low price scenarios

CO2 Emissions by emission source
LL Scenario

	1990	1995	2000	2005	2010	2015	2020
Power stations	54.3	40.6	35.3	42.6	38.2	44.0	46.6
Transport	32.9	35.5	37.5	39.4	41.6	44.8	48.4
Domestic	21.8	23.9	23.2	23.1	22.9	22.8	22.6
Services	8.3	8.7	9.0	9.4	9.9	10.5	11.1
Refineries	4.9	5.2	5.3	5.5	5.8	6.2	6.8
Other industry + agriculture	36.1	35.7	36.3	36.9	36.8	36.1	35.9
total	158.3	149.7	146.7	156.8	155.2	164.4	171.4

CO2 Emissions by emission source
CL Scenario

	1990	1995	2000	2005	2010	2015	2020
Power stations	54.3	41.2	36.1	44.1	40.3	47.5	52.1
Transport	32.9	35.6	38.0	40.6	43.6	47.8	52.5
Domestic	21.8	23.9	23.2	23.0	22.9	22.6	22.5
Services	8.3	8.8	9.2	9.6	10.3	11.1	11.6
Refineries	4.9	5.3	5.4	5.7	6.1	6.7	7.3
Other industry + agriculture	36.1	36.1	37.7	39.0	38.6	38.0	38.2
total	158.3	150.8	149.7	162.0	161.8	173.7	184.2

CO2 Emissions by emission source
HL Scenario

	1990	1995	2000	2005	2010	2015	2020
Power stations	54.3	41.8	36.6	44.8	41.9	50.1	55.2
Transport	32.9	35.7	38.6	41.8	45.6	50.7	56.4
Domestic	21.8	24.0	23.2	23.0	22.8	22.5	22.3
Services	8.3	8.8	9.3	9.9	10.7	11.5	12.1
Refineries	4.9	5.3	5.5	6.0	6.4	7.1	7.9
Other industry + agriculture	36.1	36.5	38.3	39.7	39.3	39.0	39.3
total	158.3	152.0	151.5	165.1	166.7	180.8	193.2

* as million tonnes of carbon   n.b. totals may not equal sum of parts due to rounding

Table E2 continued
High price scenarios

CO_2 Emissions by emission source
LH Scenario

	1990	1995	2000	2005	2010	2015	2020
Power stations	54.3	40.8	36.1	44.0	42.2	49.9	53.9
Transport	32.9	35.1	36.4	37.4	38.6	40.9	43.5
Domestic	21.8	23.6	22.7	22.5	22.3	22.0	21.8
Services	8.3	8.6	8.9	9.3	9.9	10.4	10.9
Refineries	4.9	5.1	5.0	5.0	5.2	5.5	5.9
Other industry + agriculture	36.1	35.2	35.1	35.5	35.5	35.9	37.1
total	158.3	148.4	144.1	153.7	153.6	164.6	173.1

CO_2 Emissions by emission source
CH Scenario

	1990	1995	2000	2005	2010	2015	2020
Power stations	54.3	41.4	38.2	45.8	44.5	54.0	60.6
Transport	32.9	35.2	36.9	38.6	40.6	43.9	47.6
Domestic	21.8	23.6	22.7	22.5	22.2	21.9	21.6
Services	8.3	8.7	9.1	9.6	10.3	11.0	11.5
Refineries	4.9	5.1	5.1	5.2	5.5	6.0	6.5
Other industry + agriculture	36.1	35.4	36.4	37.5	37.6	38.2	40.0
total	158.3	149.4	148.4	159.2	160.8	175.0	187.8

CO_2 Emissions by emission source
HH Scenario

	1990	1995	2000	2005	2010	2015	2020
Power stations	54.3	42.1	39.0	47.2	45.2	56.8	64.0
Transport	32.9	35.3	37.4	39.8	42.6	46.8	51.5
Domestic	21.8	23.6	22.7	22.5	22.1	21.8	21.5
Services	8.3	8.7	9.2	9.9	10.6	11.4	12.0
Refineries	4.9	5.1	5.2	5.5	5.8	6.4	7.1
Other industry + agriculture	36.1	36.1	37.0	38.2	38.5	39.2	41.3
total	158.3	150.9	150.5	163.0	164.9	182.4	197.4

*as million tonnes of carbon

Table E3: UK sulphur dioxide emissions by sector*

SO2 Emissions by emission source
CL Scenario

	1990	1995	2000	2005	2010	2015	2020
Power stations	2.72	1.14	0.72	0.58	0.16	0.16	0.12
Transport	0.13	0.12	0.09	0.09	0.09	0.09	0.09
Domestic	0.12	0.10	0.08	0.06	0.04	0.03	0.02
Services	0.09	0.09	0.07	0.07	0.07	0.07	0.07
Refineries	0.11	0.15	0.13	0.13	0.13	0.14	0.15
Other industry+agriculture	0.62	0.58	0.57	0.54	0.49	0.49	0.49
total	3.78	2.18	1.66	1.47	0.98	0.98	0.94

*million tonnes of sulphur dioxide

Table E4: Large Combustion Plant Directive Summary: SO2*   CL Scenario

	1990	1995	2000	2005
ESI emissions	2.722	1.140	0.720	0.580
ESI limit	2.906	2.339	1.572	1.208
Refinery emissions	0.095	0.090	0.078	0.078
Refinery limit	0.100	0.098	0.093	0.090
Other industry (OTI) emissions	0.135	0.119	0.113	0.113
Other industry limit	0.332	0.258	0.193	0.154
Total LCPD plant emissions	2.952	1.350	0.912	0.772
Total LCPD plant limit	3.338	2.695	1.858	1.452
%reduction on 1980: ESI:	-9.4	-62.1	-76.0	-80.7
%reduction on 1980: Refineries:	-64.7	-66.3	-70.8	-70.8
%reduction on 1980: OTI:	-78.3	-80.8	-81.7	-81.7
%reduction on 1980: TOTAL:	-24.2	-65.3	-76.6	-80.2

*million tonnes of sulphur dioxide

Table E5: CO2 emissions* by type of fuel

LL Scenario	1990	1995	2000	2005	2010	2015	2020
Coal	64.5	38.3	23.6	27.3	17.5	19.9	15.1
Solid smokeless fuel	2.0	2.5	2.3	2.0	1.8	1.6	1.6
Petroleum:							
motor spirit	20.8	21.3	20.9	20.0	20.1	21.1	22.4
DERV	9.1	10.9	13.3	16.1	18.1	20.1	22.1
Gasoil	7.2	7.6	6.7	6.2	5.9	5.7	5.8
Fuel oil	12.3	9.9	13.6	15.4	13.0	15.4	23.4
Burning oil	1.8	1.8	1.1	0.8	0.6	0.6	0.5
Other petroleum	4.8	4.9	4.8	4.9	5.3	5.8	6.5
Gases	29.9	46.8	54.8	58.5	67.3	68.8	68.9
Other emissions	6.0	5.8	5.7	5.6	5.7	5.3	5.2
Total	158.3	149.7	146.7	156.8	155.2	164.4	171.4

CL Scenario	1990	1995	2000	2005	2010	2015	2020
Coal	64.5	39.0	23.5	28.1	17.7	20.3	16.2
Solid smokeless fuel	2.0	2.5	2.2	2.0	1.8	1.6	1.5
Petroleum:							
motor spirit	20.8	21.3	21.1	20.0	20.3	20.6	21.9
DERV	9.1	11.0	13.6	16.8	19.4	22.1	25.0
Gasoil	7.2	7.6	6.9	6.4	6.0	5.8	5.8
Fuel oil	12.3	10.0	13.9	16.0	13.9	18.1	27.3
Burning oil	1.8	1.8	1.2	0.8	0.7	0.6	0.5
Other petroleum	4.8	4.9	5.0	5.3	5.6	6.3	7.1
Gases	29.9	47.0	56.6	61.1	70.7	73.0	73.7
Other emissions	6.0	5.8	5.7	5.6	5.7	5.3	5.2
Total	158.3	150.8	149.7	162.0	161.8	173.7	184.2

*as million tonnes of carbon   n.b. fuel oil includes orimulsion

Table E5: CO2 emissions* by type of fuel							
HL Scenario	1990	1995	2000	2005	2010	2015	2020
Coal	64.5	39.6	23.6	27.5	17.2	18.2	13.7
Solid smokeless fuel	2.0	2.5	2.2	2.0	1.7	1.6	1.5
Petroleum:							
motor spirit	20.8	21.3	21.3	20.6	21.1	22.5	24.1
DERV	9.1	11.0	13.9	17.6	20.8	24.2	27.8
Gasoil	7.2	7.6	7.0	6.5	6.1	5.9	5.8
Fuel oil	12.3	10.0	14.1	16.0	14.2	20.3	30.0
Burning oil	1.8	1.9	1.2	0.8	0.7	0.6	0.5
Other petroleum	4.8	4.9	5.1	5.6	6.0	6.7	7.7
Gases	29.9	47.3	57.4	62.9	73.2	75.6	76.9
Other emissions	6.0	5.8	5.7	5.6	5.7	5.3	5.2
Total	158.3	152.0	151.5	165.1	166.7	180.8	193.2
LH Scenario	1990	1995	2000	2005	2010	2015	2020
Coal	64.5	38.5	30.2	35.1	28.8	38.2	46.7
Solid smokeless fuel	2.0	2.4	2.1	2.0	1.8	1.7	1.8
Petroleum:							
motor spirit	20.8	21.1	20.2	18.6	18.0	18.4	19.0
DERV	9.1	10.8	13.1	15.7	17.5	19.2	21.0
Gasoil	7.2	7.1	5.5	4.6	4.1	3.8	3.6
Fuel oil	12.3	9.5	7.8	10.9	11.4	11.2	8.3
Burning oil	1.8	1.6	0.8	0.5	0.4	0.3	0.2
Other petroleum	4.8	4.7	4.4	4.4	4.7	5.1	5.7
Gases	29.9	46.9	54.2	56.3	61.3	61.3	61.6
Other emissions	6.0	5.8	5.7	5.6	5.7	5.3	5.2
Total	158.3	148.4	144.1	153.7	153.6	164.6	173.1

*as million tonnes of carbon   n.b. fuel oil includes orimulsion

Table E5: CO2 emissions* by type of fuel

CH Scenario	1990	1995	2000	2005	2010	2015	2020
Coal	64.5	39.0	32.5	35.6	29.7	40.8	50.1
Solid smokeless fuel	2.0	2.5	2.2	2.0	1.8	1.8	1.9
Petroleum:							
motor spirit	20.8	21.1	20.4	19.0	18.5	19.1	20.0
DERV	9.1	10.9	13.4	16.5	18.8	21.3	23.9
Gasoil	7.2	7.1	5.6	4.8	4.3	4.0	3.7
Fuel oil	12.3	9.6	8.1	11.6	11.8	11.7	12.1
Burning oil	1.8	1.7	0.8	0.5	0.4	0.3	0.3
Other petroleum	4.8	4.7	4.6	4.7	5.1	5.6	6.3
Gases	29.9	47.1	55.1	58.8	64.8	65.1	64.4
Other emissions	6.0	5.8	5.7	5.6	5.7	5.3	5.2
Total	158.3	149.4	148.4	159.2	160.8	175.0	187.8

HH Scenario	1990	1995	2000	2005	2010	2015	2020
Coal	64.5	40.0	33.2	36.2	28.3	42.1	49.4
Solid smokeless fuel	2.0	2.5	2.2	2.0	1.8	1.7	1.8
Petroleum:							
motor spirit	20.8	21.1	20.5	19.3	19.0	19.8	20.7
DERV	9.1	10.9	13.7	17.2	20.2	23.3	26.8
Gasoil	7.2	7.2	5.7	4.9	4.4	4.1	3.8
Fuel oil	12.3	9.7	8.3	12.1	11.4	11.6	14.8
Burning oil	1.8	1.7	0.8	0.5	0.4	0.3	0.3
Other petroleum	4.8	4.8	4.7	5.0	5.4	6.1	6.9
Gases	29.9	47.4	55.7	60.2	68.3	68.1	67.7
Other emissions	6.0	5.8	5.7	5.6	5.7	5.3	5.2
Total	158.3	150.9	150.5	163.0	164.9	182.4	197.4

*as million tonnes of carbon   n.b. fuel oil includes orimulsion

Table E6: CO2 emissions* by end user

LL Scenario	1990	1995	2000	2005	2010	2015	2020
domestic	41.7	40.1	36.6	38.4	36.4	38.0	38.7
commercial/public service	23.0	21.2	19.9	22.3	21.6	23.8	25.0
industry	48.0	42.2	42.1	45.7	44.3	45.9	47.0
agriculture	1.6	1.4	1.3	1.3	1.2	1.3	1.3
road transport	33.2	35.2	37.3	39.3	41.5	44.6	48.1
other transport	5.0	4.9	4.9	5.1	5.3	5.7	6.2
exports	1.0	1.0	1.1	1.1	1.2	1.2	1.4
other emissions	4.4	3.7	3.6	3.7	4.0	3.8	3.9
TOTAL	158.3	149.7	146.7	156.8	155.2	164.4	171.4

CL Scenario	1990	1995	2000	2005	2010	2015	2020
domestic	41.7	40.2	36.6	38.4	36.5	38.3	39.2
commercial/public service	23.0	21.3	20.2	22.9	22.5	25.1	26.7
industry	48.0	42.9	44.2	48.9	47.6	50.4	52.9
agriculture	1.6	1.4	1.2	1.3	1.2	1.3	1.3
road transport	33.2	35.2	37.8	40.5	43.5	47.6	52.2
other transport	5.0	4.9	5.0	5.3	5.5	6.0	6.6
exports	1.0	1.1	1.1	1.1	1.2	1.3	1.5
other emissions	4.4	3.7	3.6	3.7	4.0	3.8	3.9
TOTAL	158.3	150.8	149.7	162.0	161.8	173.7	184.2

HL Scenario	1990	1995	2000	2005	2010	2015	2020
domestic	41.7	40.4	36.7	38.3	36.6	38.5	39.4
commercial/public service	23.0	21.5	20.4	23.2	23.1	26.0	27.9
industry	48.0	43.8	45.1	50.3	49.4	53.0	56.1
agriculture	1.6	1.4	1.2	1.3	1.2	1.2	1.3
road transport	33.2	35.3	38.3	41.6	45.5	50.5	56.1
other transport	5.0	4.9	5.1	5.5	5.8	6.4	7.1
exports	1.0	1.1	1.1	1.2	1.3	1.4	1.6
other emissions	4.4	3.7	3.6	3.7	4.0	3.8	3.9
TOTAL	158.3	152.0	151.5	165.1	166.7	180.8	193.2

*as million tonnes of carbon

totals may not equal sum of parts due to rounding

Table E6 continued

LH Scenario	1990	1995	2000	2005	2010	2015	2020
domestic	41.7	39.8	36.4	38.3	37.1	39.1	40.0
commercial/public service	23.0	21.1	20.2	22.8	22.9	25.4	26.8
industry	48.0	41.8	40.7	44.4	43.9	47.7	50.8
agriculture	1.6	1.4	1.2	1.3	1.2	1.3	1.3
road transport	33.2	34.9	36.3	37.4	38.6	40.8	43.3
other transport	5.0	4.8	4.7	4.8	5.0	5.4	5.9
exports	1.0	1.0	1.0	1.0	1.0	1.1	1.2
other emissions	4.4	3.7	3.6	3.7	4.0	3.8	3.9
TOTAL	158.3	148.4	144.1	153.7	153.6	164.6	173.1

CH Scenario	1990	1995	2000	2005	2010	2015	2020
domestic	41.7	39.9	36.8	38.5	37.2	39.5	40.8
commercial/public service	23.0	21.3	20.9	23.5	23.8	26.8	28.9
industry	48.0	42.4	43.2	47.7	47.7	53.0	58.0
agriculture	1.6	1.4	1.2	1.3	1.2	1.3	1.3
road transport	33.2	35.0	36.8	38.6	40.6	43.8	47.4
other transport	5.0	4.8	4.8	5.0	5.3	5.7	6.3
exports	1.0	1.0	1.0	1.0	1.1	1.2	1.3
other emissions	4.4	3.7	3.6	3.7	4.0	3.8	3.9
TOTAL	158.3	149.4	148.4	159.2	160.8	175.0	187.8

HH Scenario	1990	1995	2000	2005	2010	2015	2020
domestic	41.7	40.1	36.9	38.6	37.0	39.7	41.0
commercial/public service	23.0	21.5	21.2	24.0	24.2	27.8	30.0
industry	48.0	43.5	44.2	49.3	49.4	55.9	61.8
agriculture	1.6	1.4	1.2	1.3	1.2	1.3	1.3
road transport	33.2	35.0	37.3	39.7	42.6	46.7	51.3
other transport	5.0	4.8	4.9	5.3	5.5	6.1	6.7
exports	1.0	1.0	1.0	1.1	1.2	1.3	1.4
other emissions	4.4	3.7	3.6	3.7	4.0	3.8	3.9
TOTAL	158.3	150.9	150.5	163.0	164.9	182.4	197.4

*as million tonnes of carbon

totals may not equal sum of parts due to rounding

ANNEX F

ENERGY DEMAND AND EFFICIENCY TRENDS IN THE INDUSTRIAL SUB-SECTORS

This annex gives a more detailed insight into the energy demands and energy ratios associated with the different industrial sub-sectors. An energy ratio is a crude measure of energy efficiency, showing energy demand per unit of output. A fall in a sector's energy ratio can imply an increase in the efficiency with which it uses energy. Increased efficiency can arise from technical improvements (e.g. machinery becomes more efficient in its use of energy), from fuel switching (away from less efficient fuels such as oil and coal, into gas or electricity), or from a changing mix of output within the sub-sector. Changes in output can also affect the energy ratio, for instance increased output is likely to lead to a fall in the energy ratio as fixed energy overheads form a smaller proportion of a sector's energy demand.

For each of the following eight sub-sectors the energy ratio in 1990 has been compared with the projected 2000 and 2020 ratios for all six main scenarios:

- Iron and Steel (I&S)
- Non-Ferrous Metals (NFM)
- Mineral Products (MIN)
- Chemicals (CHM)
- Engineering and Vehicles (ENG)
- Food, Drink and Tobacco (FDT)
- Textiles, Leather and Clothing (TLC)
- Paper, Printing and Publishing (PPP)

Tables F.1, F.2 and F.3 show the EP65 energy ratios for the central, high and low GDP growth scenarios respectively. The ratios have been calculated on the basis of *delivered* energy demand relative to output.

TABLE F.1 - Central GDP Growth Industrial Sector Energy Ratios						
	Low Fuel Prices			High Fuel Prices		
	1990	2000	2020	1990	2000	2020
I&S	1.00	0.90	0.89	1.00	0.91	0.88
NFM	1.00	1.03	0.86	1.00	1.02	0.87
MIN	1.00	0.60	0.42	1.00	0.58	0.40
CHM	1.00	0.81	0.44	1.00	0.75	0.41
ENG	1.00	0.86	0.69	1.00	0.80	0.65
FDT	1.00	0.99	0.83	1.00	0.94	0.79
TLC	1.00	1.28	1.02	1.00	1.14	0.89
PPP	1.00	0.88	0.68	1.00	0.80	0.63

TABLE F.2 - High GDP Growth Industrial Sector Energy Ratios						
	Low Fuel Prices			High Fuel Prices		
	1990	2000	2020	1990	2000	2020
I&S	1.00	0.90	0.91	1.00	0.91	0.89
NFM	1.00	0.97	0.76	1.00	0.96	0.78
MIN	1.00	0.60	0.40	1.00	0.57	0.38
CHM	1.00	0.80	0.41	1.00	0.75	0.38
ENG	1.00	0.85	0.66	1.00	0.80	0.63
FDT	1.00	0.97	0.77	1.00	0.92	0.74
TLC	1.00	1.28	0.99	1.00	1.15	0.90
PPP	1.00	0.88	0.63	1.00	0.79	0.60

TABLE F.3 - Low GDP Growth Industrial Sector Energy Ratios						
	Low Fuel Prices			High Fuel Prices		
	1990	2000	2020	1990	2000	2020
I&S	1.00	0.89	0.84	1.00	0.90	0.83
NFM	1.00	1.08	0.95	1.00	1.07	0.97
MIN	1.00	0.62	0.44	1.00	0.59	0.42
CHM	1.00	0.84	0.49	1.00	0.78	0.44
ENG	1.00	0.92	0.75	1.00	0.86	0.71
FDT	1.00	0.99	0.89	1.00	0.94	0.83
TLC	1.00	1.27	0.99	1.00	1.12	0.81
PPP	1.00	0.89	0.73	1.00	0.80	0.66

It can be seen from the tables that, with one or two exceptions, the ratios continue to decline over time. There are, however, differences in the rate at which this decline takes place and exceptions to the trend; between 1990 and 2000 in the textiles, leather and clothing sector and, in some scenarios, the non-ferrous metals sector display an increase in their energy ratios. In the non-ferrous metals sector this is because output declines between 1990 and 2000 in the low GDP scenario, resulting in fixed energy demand overheads being spread over smaller output. The increase in the textiles, leather and clothing sector's energy ratio is due to an increase in demand between 1992 and 1993 (see table 10, page 37 of the 1994 Digest of United Kingdom Energy Statistics)[1]. A more useful comparison is therefore between 2000 and 2020.

The mineral products sector has by far the largest decline (38%-43%) in its energy ratio between 1990 and 2000. The magnitude of this decline can best be explained by looking at trends in the mineral products sector's energy ratio in the early 1990s. In the 20 years prior to 1990 the energy ratio for this sector fell by an average of about 2% per annum, with fairly constant output. However, between 1990 and 1993 the ratio fell by 21%, an average of 7% per annum. This large fall in the energy ratio occurred in the presence of declining output (a fall in output would usually be associated with an increase in the energy ratio). The ratio is projected to fall by 24%-29% between 1993 and 2000 (an average of 3.4%-4.1% per annum). The continued decline of the sector's energy ratio, at a higher rate than its long-run average of 2% per annum, can be attributed to the assumed growth in output. This higher output leads to higher load factors being used on existing fixed capital stock (more energy efficient) and encourages investment in new, more efficient capital.

The long-run (between 1990 and 2020) fall in the energy ratio is also highest in the mineral products sector (56%-62%). However, the chemicals sector energy ratio is projected to fall by a very similar amount over this longer period (51%-62%). This is partly related to the effect that high output growth has in reducing the average age of the capital stock in the chemicals sector.

---

[1] There is some doubt about the accuracy of the textiles, leather and clothing sector's petroleum consumption recorded in the 1994 DUKES. This may be distorting the sector's 1995 energy ratio.

In general the energy ratio is lower in the high fuel price scenarios than their low fuel price equivalents. This is because higher fuel prices provide a large incentive for taking measures to improve efficiency. However, in the iron and steel and the non-ferrous metals sectors the ratio is very similar in the high and low fuel price scenarios.

In the non-ferrous metals sector a switch from gas into solid fuels in the high price scenarios leads to an unexpectedly high energy ratio. The price of solid fuels rises by less than the prices of other fossil fuels in the high price scenarios and it becomes worthwhile for the non-ferrous metals sector to buy more delivered therms of solid fuel than the smaller amount of gas needed to satisfy a given energy demand. This pushes up the energy ratio in the high price scenarios.

In the iron and steel sector, a fairly constant level of output across all scenarios (beyond 2005) and little projected fuel switching leads to an energy ratio that does not alter much across scenarios.

As would be expected, the energy ratios are lower in the higher GDP growth scenarios. Increased output leads to fixed energy overheads being spread over higher output and the energy ratio therefore falls.

The general downward trend in the industrial sector energy ratios should be treated with caution, as the decline is often partly due to switching from less efficient fossil fuels into electricity, which has a very high efficiency at point of use, but which is typically produced from power stations with relatively low thermal efficiencies.

In summary, the projections contain continuing improvements in the sub-sectoral energy ratios within industry. When the longer term is considered, 2000 - 2020, it can be seen that, with the exception of the iron and steel and chemicals sectors, an improvement of approximately 1% per annum occurs in each sector's energy ratio. The iron and steel energy ratio remains fairly constant over this period, while the chemicals sector ratio improves by about 2% per annum.

ANNEX G

SENSITIVITY OF EP65 PROJECTIONS TO VERY HIGH AND VERY LOW FUEL PRICES

This annex looks at projected final user energy demand, CO_2 emissions and primary energy demand with very high and very low fuel prices in the short-run (up to and including 2000). The final user energy demands by sector in the extreme scenarios, high GDP growth - very low fuel prices (HVL) and low GDP growth - very high fuel prices (LVH), are shown in Tables G.1 and G.2 respectively:

TABLE G.1 - FINAL ENERGY DEMAND BY SECTOR (HVL)			
			(Billion Therms)
	1990	1995	2000
Domestic	16.2	18.1	18.3
Iron & Steel	3.0	2.9	3.3
Other Industry	12.2	12.2	13.7
Services	7.8	8.2	8.9
Transport	19.3	21.3	24.3
Total	58.4	62.7	68.6

TABLE G.2 - FINAL ENERGY DEMAND BY SECTOR (LVH)			
			(Billion Therms)
	1990	1995	2000
Domestic	16.2	17.7	17.4
Iron & Steel	3.0	2.7	3.0
Other Industry	12.2	11.2	11.0
Services	7.8	8.1	8.7
Transport	19.3	20.4	19.9
Total	58.4	60.1	60.0

Table G3, overleaf, shows the projected levels of carbon dioxide and primary energy demand in 2000 for the two extreme scenarios (HVL and LVH) and the two closest 'normal' scenarios (HL and LH).

TABLE G3 - THE RANGE OF CARBON DIOXIDE EMISSIONS AND PRIMARY DEMAND IN 2000		
	carbon dioxide (MtC)	primary energy demand (Mtoe)
High growth, very low energy prices (HVL)	154.7	245.1
HL scenario	151.5	239.9
LH scenario	144.1	226.0
Low growth, very high energy prices (LVH)	136.5	216.3

ANNEX H

FINAL USER ENERGY DEMAND PRICE ELASTICITIES

This annex provides estimates of the DTI energy model's price elasticities. The price elasticity of demand is a measure of the responsiveness of energy demand to changes in price. It is formally defined as:

$$\frac{\text{percentage change in energy demand}}{\text{percentage change in energy price}}$$

Elasticities in the model are not constants but vary across time. The elasticity values are also dependent on factors such as the size of the change in price, the existing price level, the timing of the price change and the level of energy efficiency in the existing capital stock. Thus the elasticities shown in Table H.1 are meant to give only a general impression of the typical values of the elasticities.

Price elasticities are normally negative, as an increase in the level of energy prices encourages consumers to use less energy in total and to use the energy that they do consume more efficiently. Thus the more negative an energy price elasticity is, the more responsive energy demand is to energy price increases. Since consumers have more limited possibilities for reducing their energy consumption in the short term than in the long term, energy price elasticities are usually shown for both time periods.

Table H.1	Short-Run	Long-Run
Domestic sector	-0.04	-0.19
Iron & Steel sector @	-0.06	-0.07
Non-Ferrous Metals	-0.10	-0.15
Mineral Products	-0.08	-0.16
Chemicals	-0.10	-0.41
Engineering & Vehicles	-0.25	-0.56
Food, Drink & Tobacco	-0.14	-0.29
Textiles, Leather & Clothing	-0.36	-0.45
Paper, Printing & Publishing	-0.31	-0.42
Construction and Other Industries	-0.11	-0.36
TOTAL OTHER INDUSTRIES	**-0.17**	**-0.38**
Service sector	-0.06	-0.09
Road Transport sector	-0.12	-0.41

@ refers to coke and secondary gases.

In order to understand the elasticities shown in Table H.1 some examples may be helpful. Thus a 10% increase in energy prices in the domestic sector would result in a 1.9% reduction in energy demand in the long run. A similar 10% increase in energy prices in the other industries would result in a 3.8% reduction in energy demand in the long run.

The elasticities shown in Table H.1 are for aggregate energy demand (i.e. the sum of electricity, gas, oil and solid fuel demands) and therefore incorporate any inter-fuel switching that may occur as a result of the increase in energy prices. The price elasticity for any given fuel may therefore be somewhat larger than the aggregate price elasticities shown in Table H.1. It would be misleading therefore to apply the price elasticities shown above to a single fuel in any particular sector.

ANNEX I

ENERGY DEMAND MODEL SCHEMATICS

This annex contains schematics which show the general structure of the main final user energy demand models.

Diagram I.1	Domestic sector model diagram
Diagram I.2	Service sector model diagram
Diagram I.3	Other Industry sector model diagram
Diagram I.4	Road Transport sector model diagram
Diagram I.5	Iron and Steel sector process flow diagram

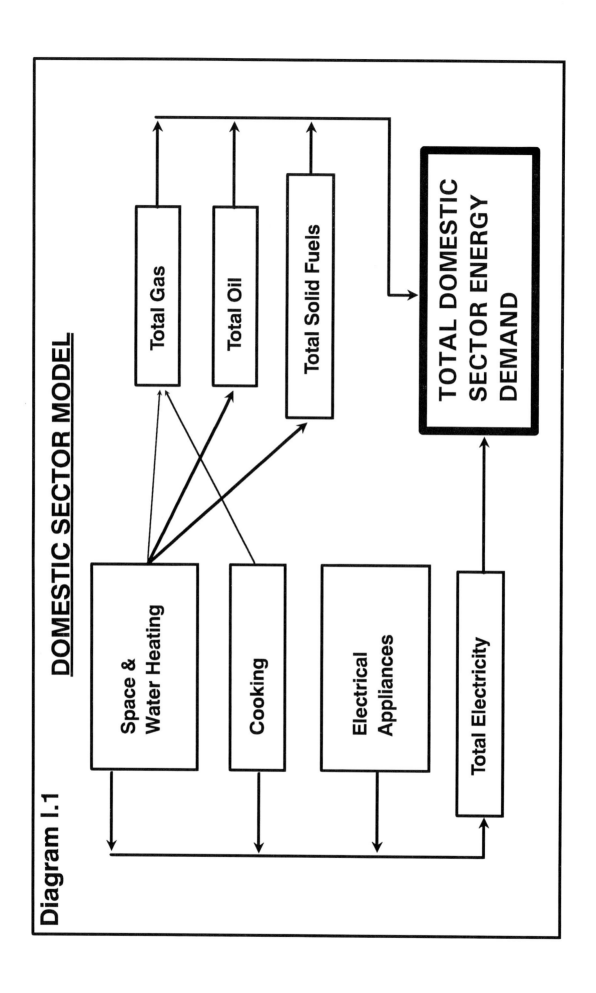

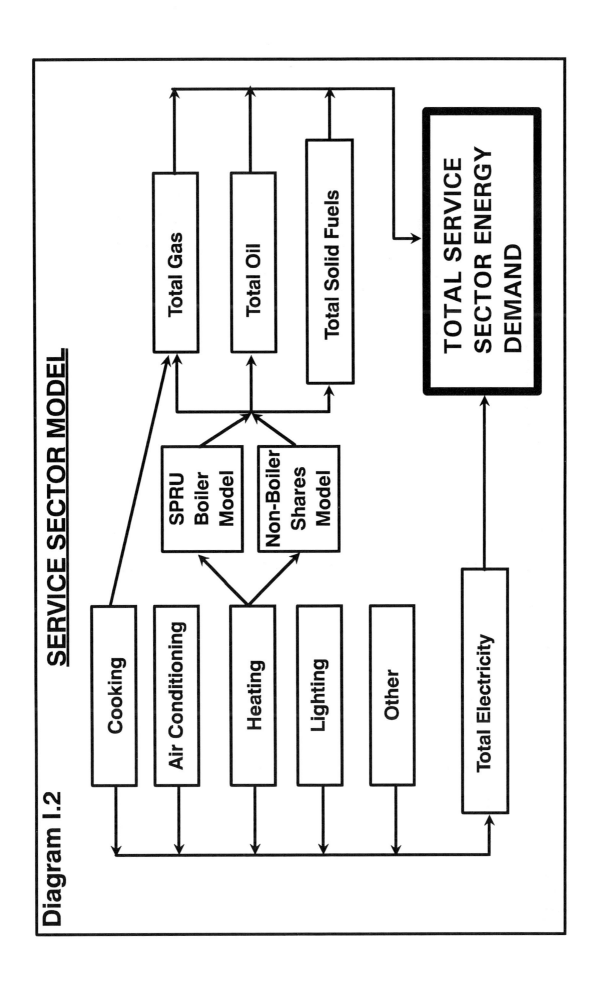

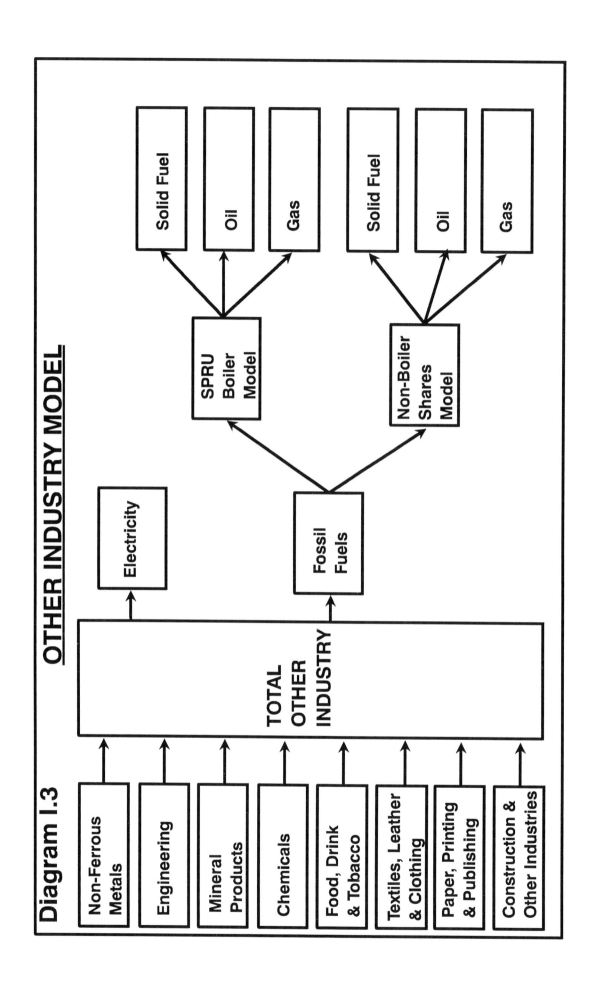

Diagram I.4

ROAD TRANSPORT ENERGY DEMAND: MODEL STRUCTURE

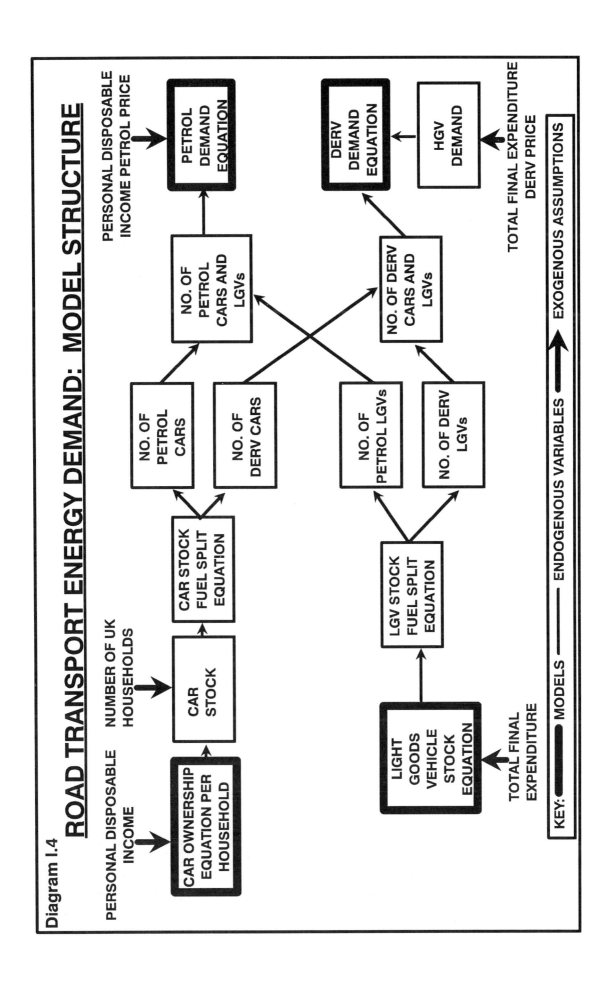

Diagram I.5　Simplified iron and steel manufacturing process flow diagram

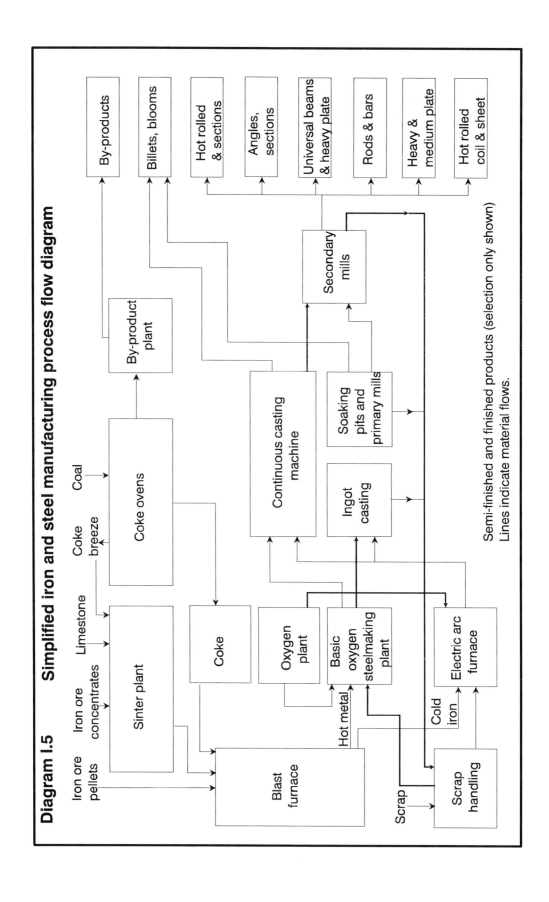

ANNEX J

GLOSSARY

ARA	Amsterdam, Rotterdam, Antwerp
BOS	basic oxygen steel
BRE	Building Research Establishment
CCGT	combined cycle gas turbine
CCP	Climate Change Programme
CH	central GDP growth - high fuel prices
CHP	combined heat and power
CL	central GDP growth - low fuel prices
CO_2	carbon dioxide
DERV	diesel engine road vehicle
DoE	Department of Environment
DTI	Department of Trade and Industry
DUKES	Digest of UK Energy Statistics
EAF	electric arc furnace
EC	European Community
EEO	Energy Efficiency Office
EP58	Energy Paper 58
EP59	Energy Paper 59
EP65	Energy Paper 65
ESI	electricity supply industry
ETSU	Energy Technology Support Unit
FCCC	Framework Convention on Climate Change
FGD	flue gas desulphurisation
FSU	Former Soviet Union
GCI	gas cost index
GDP	gross domestic product
GW	gigawatts
HDGT	heavy duty gas turbine
HH	high GDP growth - high fuel prices
HL	high GDP growth - low fuel prices
HMIP	Her Majesty's Inspectorate of Pollution
HMT	Her Majesty's Treasury
IEA	International Energy Agency
IGCC	integrated gasification combined cycle
KW	kilowatts
LCPD	Large Combustion Plant Directive

LH	low GDP growth - high fuel prices
LL	low GDP growth - low fuel prices
LNG	liquefied natural gas
MMC	Monopolies and Merger Commission
MtC	million tonnes of carbon
MW	megawatts
NAEI	National Atmospheric Emissions Inventory
NFFO	non-fossil fuel obligation
NOx	nitrogen oxides
OEF	Oxford Economic Forecasting
OFGAS	The Office of Gas Supply
PWR	pressurised water reactor
RECs	Regional Electricity Companies
RPI	retail price index
SAVE	Specific Actions for Vigorous Energy efficiency
SO_2	sulphur dioxide
SEEC	Surrey Economics Energy Centre
SPRU	Science Policy Research Unit
toe	tonnes of oil equivalent
UNECE	United Nations Economic Commission for Europe
VAT	value added tax
WACOG	weighted average cost of gas

MARKLAND LIBRARY   BECK LIBRARY
Tel. 737 3528   Tel.
Telephone Renewals can only
be accepted after 4.30 p.m.

164351

This book is to be returned on or before
the last date stamped below.

2 5 APR 1997

LIBREX

DTI.   164351

LIVERPOOL HOPE
THE BECK LIBRARY
HOPE PARK, LIVERPOOL, L16 9JD